Intelligent Systems Reference Library

Volume 240

The aim of this series is to publish a Reference Library, including novel advances and developments in all aspects of Intelligent Systems in an easily accessible and well structured form. The series includes reference works, handbooks, compendia, textbooks, well-structured monographs, dictionaries, and encyclopedias. It contains well integrated knowledge and current information in the field of Intelligent Systems. The series covers the theory, applications, and design methods of Intelligent Systems. Virtually all disciplines such as engineering, computer science, avionics, business, e-commerce, environment, healthcare, physics and life science are included. The list of topics spans all the areas of modern intelligent systems such as: Ambient intelligence, Computational intelligence, Social intelligence, Computational neuroscience, Artificial life, Virtual society, Cognitive systems, DNA and immunity-based systems, e-Learning and teaching, Human-centred computing and Machine ethics, Intelligent control, Intelligent data analysis, Knowledge-based paradigms, Knowledge management, Intelligent agents, Intelligent decision making, Intelligent network security, Interactive entertainment, Learning paradigms, Recommender systems, Robotics and Mechatronics including human-machine teaming, Self-organizing and adaptive systems, Soft computing including Neural systems, Fuzzy systems, Evolutionary computing and the Fusion of these paradigms, Perception and Vision, Web intelligence and Multimedia.

Indexed by SCOPUS, DBLP, zbMATH, SCImago.

All books published in the series are submitted for consideration in Web of Science.

Tushar Bhardwaj · Himanshu Upadhyay ·
Tarun Kumar Sharma · Steven Lawrence Fernandes
Editors

Artificial Intelligence in Cyber Security: Theories and Applications

Springer

Editors
Tushar Bhardwaj
Applied Research Center
Florida International University
Miami Fl, FL, USA

Himanshu Upadhyay
Applied Research Center
Florida International University
Miami Fl, FL, USA

Tarun Kumar Sharma
Shobhit University
Gangoh, India

Steven Lawrence Fernandes
Department of Computer Science
Creighton University
Omaha, NE, USA

ISSN 1868-4394 ISSN 1868-4408 (electronic)
Intelligent Systems Reference Library
ISBN 978-3-031-28580-6 ISBN 978-3-031-28581-3 (eBook)
https://doi.org/10.1007/978-3-031-28581-3

This Springer imprint is published by the registered company Springer Nature Switzerland AG
The registered company address is: Gewerbestrasse 11, 6330 Cham, Switzerland

Preface

This book highlights the applications and theory of artificial intelligence in the domain of cyber security. The idea is to propose new approaches and ideas, and to present applications of innovative approaches in real-time environments. In the past few decades there has been an exponential rise in the application of artificial intelligence technologies (such as deep learning, machine learning, blockchain, etc.) for solving complex and intricate problems arising in the domain of cyber security. The versatility of these techniques has made them a favorite among scientists and researchers working in diverse areas.

The book comprises of **9 Chapters**. In **Chap. 1**, the authors provide a comprehensive and systematic literature review on the use of ML in cloud security and its applications and techniques to prevent security issues on cloud computing. **Chapter 2** introduces the heterogeneous model for WSNs and discusses the clustering techniques for HWSN networks. Also authors have conducted an exhaustive survey of the heterogeneous network in wireless sensor models and discussed about the top priority being granted to productive energy usage. **Chapter 3** compares various sentimental analysis techniques to classify the restaurant reviews as positive or negative sentiments to find the most suitable one to use to achieve maximum accuracy in terms of classification between positive or negative reviews and also automated the approach. **Chapter 4** presents many applications of machine learning and cyber security that are currently employed in various industries. **Chapter 5** presents various challenges that arise when machine learning is integrated with cybersecurity. **Chapter 6** discusses the need for detecting anomalies in thermal power plant data for monitoring and maintaining the conditions of these power stations and discusses various artificial intelligence (AI) techniques to detect sensor data anomalies in different components of a thermal power plant. **Chapter 7** analyzes deep learning (DL) models, formed with three hidden layers and three different optimizer functions for intrusion detection in the data traffic. In **Chap. 8**, the authors propose a machine learning-based framework with various supervised learning algorithms such as Random Forest, Support Vector Machine, and Decision Trees. The authors perform the hyper-parameter optimization of these algorithms using sci-kit-learn machine learning frameworks. In **Chap. 9**, the authors address the problem of analyzing fraudulent credit card transactions. Explicitly, we propose a deep learning-based framework with various unsupervised learning algorithms. The authors have performed the hyperparameter optimization of these algorithms using sci-kit-learn machine learning frameworks and popular deep learning framework TensorFlow.

We hope that the effort will be found informative and interesting to those who are keen to learn on technologies that address to the challenges of the exponentially growing information in the core and allied fields of cyber security. This book will be beneficial to the young scholars and researchers working in the allied domain. We are thankful to the authors of the chapters for their valuable contribution to the conference and for bringing forth significant research and literature across the field of cyber security. Offering valuable insights into cyber security for the young scholars, researchers, academicians and industrialists, this book will inspire further research in this dynamic field. We express special thanks to reviewers, SpringerNature and team for their valuable support in the publication of the edited book.

Miami Fl, USA Tushar Bhardwaj
Miami Fl, USA Himanshu Upadhyay
Gangoh, India Tarun Kumar Sharma
Omaha, USA Steven Lawrence Fernandes

Contents

Chapter 1
A Comprehensive Review on Cloud Security Using Machine Learning Techniques

Divya Gangwani, Harshal A. Sanghvi, Viral Parmar, Riki H. Patel, and Abhijit S. Pandya

Abstract With the increasing demand and popularity in the usage of cloud computing, there has been a necessity to prevent common attacks and security threats to cloud computing services. The consumers of cloud services are constantly concerned about the cyber-security risks, data loss, and slowdown of services. With the advancement of machine learning techniques, learning-based methods for security applications are gaining tremendous popularity in the field of literature. Over the past few years, ML techniques have been shown to prevent as well as to detect security attacks on the cloud. In this paper, we provide a comprehensive and systematic literature review on the use of ML in cloud security and its applications and techniques to prevent security issues on cloud computing. We further evaluated relevant research and studies and divide them into three main categories: (1) The security threats and attacks on cloud computing, (2) Types of ML technologies used to prevent security threats, (3) Evaluating the results and discussing the performance outcome of the models. The extensive review and findings proposed in this paper can contribute to further enhancements and improvements in cloud computing and security issues with the use of machine learning algorithms. Moreover, it will provide the basis for other researchers to contribute new ideas and enhancements in achieving reliable and safe methods to access cloud computing applications and avoid any potential security issues.

Keywords Cloud computing · Cyber security · Machine learning · Security · Threats

D. Gangwani (✉) · H. A. Sanghvi · R. H. Patel · A. S. Pandya
Department of Computer Science and Engineering, Florida Atlantic University, Boca Raton, USA
e-mail: dgangwani2017@fau.edu

H. A. Sanghvi
e-mail: hsanghvi2020@fau.edu

V. Parmar
School of Technology, Pandit Deendayal Energy University, Gandhinagar, Gujarat, India

T. Bhardwaj et al. (eds.), *Artificial Intelligence in Cyber Security: Theories and Applications*, Intelligent Systems Reference Library 240,
https://doi.org/10.1007/978-3-031-28581-3_1

1.1 Introduction

Nowadays, most of the communication with people is done over the internet. The internet is used in many of the fields in the world including medical, electronics, military, private companies, industries, education, etc. Without the internet, the world would be at a standstill. Hence, cloud computing has recently become very famous due to the increasing demand to provide more efficient and secured ways of communication and storing big data in the network. Due to the increasing computational cost, there is a growing financial burden which is caused due to storage and analysis performed on the data that imposes important modifications performed on the cloud computing to data [1]. Cloud computing refers to the on-demand availability of end-user resources, including data storage and processing capabilities without requiring the active involvement of the client. It means that the end-user can request the desired number of resources required to cater to the request of the end-user.

Figure 1.1 shows the service models of Cloud Computing. Cloud computing is a distributed computing platform that delivers a variety of services to the end-user over the internet. These services can be of three types (1) Software as a Service (SaaS), (2) Platform as a Service (PaaS), and (3) Infrastructure as a Service (IaaS).

In terms of SaaS, PaaS, and IaaS, the service provides software, platforms, and infrastructures based on the needs of the customer. The user's business data, which includes personal information and business rules, is uploaded to all the services. Personal information and business rules must be protected as this improves user trust and reliability. As a result, security in SaaS, PaaS, and IaaS is very critical [2]. The

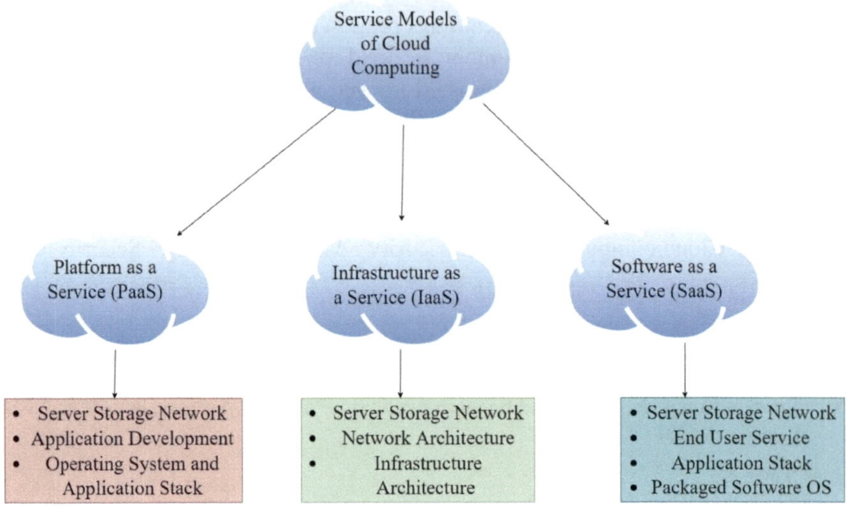

Fig. 1.1 Service models of cloud computing

current security level is very low not only in terms of services, but also in terms of unauthorized users, access permission, connection or route, data, and data storage.

An important and notable feature of cloud computing is that it is highly elastic, giving enterprises the ability to either increase or decrease the resources based on demand which results in significant cost savings. Hence, there is a need for an accurate and user-friendly model that can support large applications of cloud services and infrastructure. Currently, there are many challenges faced by cloud computing models rather than providing advantages, because every customer has different demands or needs different services. With this scenario, the problem of security rises. The architecture of the cloud faces many security challenges and according to the study, cloud security is ranked as one of the major issues which need immediate attention. Each log file generated by the connected computers to the users contains information such as IP address, host, and network protocols. These log files can be used as an investigating starting point to audit security challenges and provide improvement strategies. However, the size of the log file increases with time and becomes too huge over time. Moreover, the log files are unstructured and need intervention to address information from such huge files. Hence there is a need for big data algorithms to analyze these files and generate results [3]. Nowadays, using Machine Learning (ML) algorithms [4] provides significant results for predictive modeling. ML has the power to take huge data files and make them machine-readable to provide prediction results. This paper focuses on highlighting the use of ML to solve security and threats associated with cloud computing and provides a detailed study of how to overcome challenges faced in CC.

ML algorithms are responsible for solving security-related problems and can handle large datasets more efficiently. ML models [5] focus on the use of finding interesting patterns and relations by observing the data and channelizing the information to predict and make better decisions to solve the security issues and threats in CC and improve the performance of the model. Our main objective is to provide sufficient literature on the cloud framework focusing on the security and threats related to CC and demonstrate various types of ML algorithms that provide a good solution to overcome the issue of cloud security, challenges, and risks associated with it.

The rest of this paper is organized as follows: In Sect. 1.1, we provide a review of the related work on cloud security using machine learning applications. In Sect. 1.2, we introduce the cloud computing framework and provide the architecture of CC. The security threats and attacks on cloud services are discussed in Sect. 1.3. Section 1.4 highlights the types of ML algorithms suitable for overcoming security-related issues in CC and finally, we conclude the paper in Sect. 1.5.

1.2 Related Work

In this section, we cover the detailed analysis and provide relevant studies on security and threats related to cloud computing. Recent studies have used ML approaches to evaluate the security issues in the cloud environment and provide reasonable solutions to those problems. Classification and regression algorithms have shown tremendous improvement in many studies. In [6], The author demonstrated the use of an SVM classifier to enhance cloud and network security issues. The advantage of using SVM is that it does not rely on human judgment to make final decisions, hence SVM is one of the most reliable tools for extracting information from a linear classification model. Gordon et al. [1] highlights the detailed study on hybrid cloud computing techniques and demonstrate the ways of security and risk management at the enterprise level. A recent study [7] proposed the use of a deep learning algorithm for the detection of anomalies in cloud services. A detailed study has been carried out using deep learning approaches [8] and the performance of different models has been compared based on the accuracy.

Other studies [9, 10], proposed the use of a secured cloud management log based on temporal data that records the history and transactional data at various time points to provide security to the customers using encryption services for the log data within a particular period as requested by the user.

Decision Tree classifier and regression techniques [7] have also been most commonly used to explore different possibilities of enhancing cloud security in various environments. Supervised ML algorithms have been successfully used in cloud security and threat detection based on the given transactional records. Another ML technique, Naïve Bayes algorithms have been used in [11] to detect threats related to cloud network and achieved good results.

ML algorithms have also been used in malware detection and security threats in CC. Selamat et al. [12] provide a study that uses three ML algorithms to identify and examine the data to determine the malware issues in CC and then selects the best model based on the accuracy achieved. With the advances in ML algorithms to identify and detect the threats in computing technologies, there has been a rise in attackers as well such as the Denial of Service (DoS) attacks in Distributed DoS (DDoS) systems. The DDoS attacks have been identified as one of the major threats in the cloud computing system.

The previous research mentioned in this section, uses either one, two- or three-ML algorithms to solve cloud security issues based on the type of information available in the database. Our research area focuses on discussing different types of ML algorithms and how we can benefit from using those ML algorithms in different areas of cloud security.

1.3 Cloud Computing

When you think of cloud computing today, it has always been a long-awaited dream or a vision that has been foreseen by many people. Cloud computing (CC) is a virtualization system in which users or data owners can store and access their data remotely from any part of the world. With this technology, the users can be free of the burden to store and manage their data locally by using the shared pool of programmable sources. With the recent advances, CC can provide innovative and technologically compatible services and products to the vast spread of crowd as and when needed with different pricing options. It does so by pooling a variety of resources together to ensure immediate access and availability of resources to the users depending on the demand of services. The cloud provider ensures that resources are utilized promptly without causing any delay as it provides flexibility and a fit-for-purpose solution to the end-users. Hence, nowadays, CC is being utilized in almost every part of the business or situation and the demand has also increased with time as more and more users are requiring cloud services for their professional or personal use [13].

Nonetheless, using the same pool of resources by multiple users who share the same network can cause risks involving danger to the sensitive information being transferred on the same network. Even if a trusted person who has access to the sensitive information can access, transfer and edit the data can cause potential fraud including resource elimination, feature deletion, data leak, disabling the access and even deleting the traces of such information loss [3].

As a result, the cloud architecture has some flaws which need to be taken care of when accessing vital information. Every day, the enterprises and the users face new attacks and threats that have not yet been taken care of. Hence cloud security is a major topic that needs more attention today as the seriousness of providing security to the enterprises is vital and there is a need for improvement in cloud security to improve potential users' demand and provide a better adoption rate in the future.

1.4 The Architecture of Cloud Services

The architecture of the Cloud is generally categorized into Private, Public, and Hybrid Cloud.

Public Cloud: Computing offerings provided with the aid of using an agency that allows you to create a digital example and to apply different cloud services and products online is referred to as a public cloud. Any agency first builds its non-public cloud and after sure years of exercise with a non-public cloud agency sooner or later gives its cloud-primarily based offer.

Private Cloud: Computing offerings are restrained to a Specific business enterprise or institution of businesses to create a digital example and to apply different cloud services and products online is known as non-public cloud. This form of organization or business enterprise maintains its offerings restrained to its very own users.

Hybrid Cloud: Hybrid cloud is an answer that mixes a personal cloud with one or extra public cloud offerings, with proprietary software programs allowing communique among every awesome service. A hybrid cloud approach gives corporations more flexibility through shifting workloads among cloud answers as desires and charges fluctuate. Hybrid cloud offerings are effective due to the fact they provide corporations more manage over their facts.

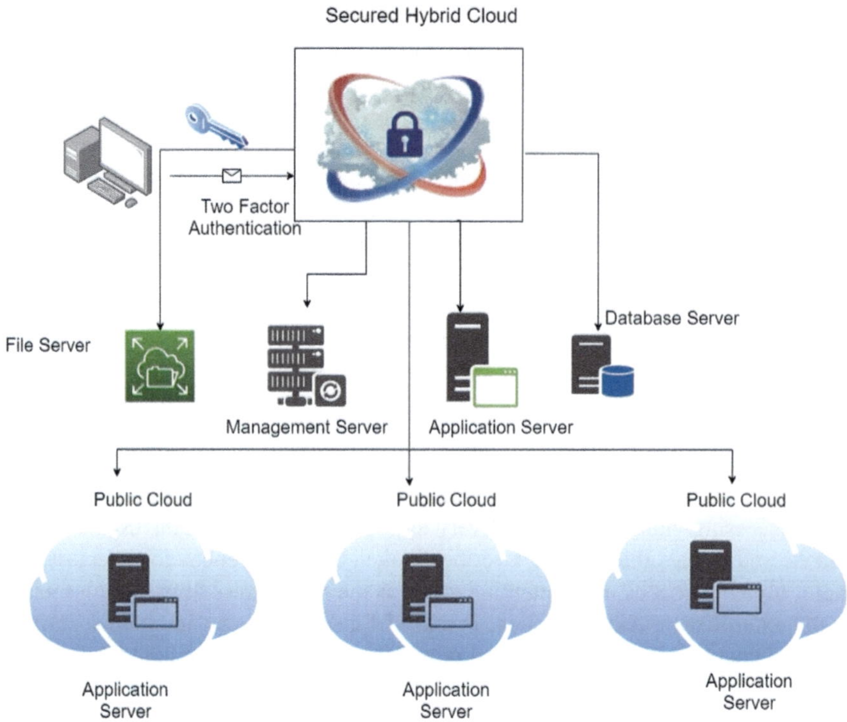

Fig. 1.2 Architecture of hybrid cloud

Figure 1.2 shows the diagram of the architecture of the hybrid cloud. The hybrid cloud is accessed through the two-factor authentication method to make it more secure. In hybrid cloud operations File Servers, Management Servers, Application Servers, and Database Servers are located. Further, the hybrid cloud is utilized as a public cloud as required by the infrastructure in the company. The application-based server manages the public cloud in the company.

1.5 Cloud Security and Threats

1.5.1 Introduction to Cloud Security

Cloud safety is a fixed set of guidelines and tactics that assist industries to defend their cloud-primarily based totally on structures and facts. These safety features are designed to deal with diverse safety concerns, which include protecting consumer privacy and making sure regulatory compliance with the proper cloud safety solution. Various companies personalize their guidelines and tactics to deal with commercial enterprise needs. This gets rid of the want for guide paintings and enables reduced costs. The manner cloud safety is introduced relies upon the kind of cloud provider that the commercial enterprise makes use of and the safety answers which might be in place.

Cloud security is a comprehensive set of technologies, protocols, and best practices to protect cloud infrastructure, cloud applications, and data hosted by the cloud. The security of cloud services begins with understanding what is secure and which device components need to be maintained. For example, cloud providers are responsible for backend development to deal with safety violations. In addition to choosing a security-conscious provider, customers should be fully informed of the appropriate settings of the vendor and safe usage practices. Customers must also ensure the security of all end-user devices and networks.

1.5.2 Segmentation of Cloud Security Responsibilities

Cloud companies are trying to create their customers a secure cloud. Their business approach focuses on preventing violations and maintaining public and consumer confidence. Cloud service providers can try to avoid cloud security risks in the service they provide, but they have no control over how customers use the service or what data they add to it. or how they have access to the services they have. Customers can compromise the security of the cloud network by setting up sensitive data and enforcing access controls. Cloud service providers and cloud service customers share different levels of security responsibilities for different types of public cloud services. By service type:

- Software as a Service (SaaS) Customer is responsible for user data and access security.
- The customer of the Platform as a Service (PaaS) is responsible for data, user access, and application security.
- Customers of Infrastructure as a Service (IaaS) are responsible for protecting data, user access, applications, operating systems, and virtual network traffic.

Customers are charged with protecting their data and controlling who has access to it in any type of public cloud service. Cloud data security is the key to successfully adopting and reaping the benefits of cloud computing. Organizations considering popular SaaS offerings such as Microsoft Office 365 or Salesforce should consider how they will meet their shared responsibility for the security of cloud data. Companies looking at IaaS services like Amazon Web-based Services (AWS) or Microsoft Azure start with data but require a more comprehensive plan that includes cloud application security, operating system security, and virtual network traffic, all of which can be problematic.

1.5.3 Cloud Security Challenges

Certain policies can keep the cloud safe because public cloud data is stored by third parties and has access to the internet.

Cloud data visibility—In most cases, cloud services are available outside the corporate network and on unprocessed devices. In other words, unlike traditional network traffic monitoring tools, the IT command requires the ability to ensure that the cloud service has full visibility of the data.

Multiagency—publicly available cloud environments concentrate multiple client infrastructures under the same umbrella, allowing hosted services to corrupt malicious attackers to ensure damage to onboard damage to other companies.

Cloud data control—When it comes to controlling the server and application in its configuration, a third-party cloud service provider's IT team is small. Cloud customers have limited default controls and cannot access basic physical infrastructure.

Access to cloud data and applications—Studios can now access cloud applications and data over the Internet, making access control based on traditional network data center boundaries ineffective. Moreover, privileged access to cloud-providing staff may bypass its safety controls.

Compliance—Using cloud-based services adds a new dimension to regulatory and internal compliance. Regulations such as HIPAA, PCI, and internal teams, partners, and customers may require cloud environment compliance. Compliance and risk management include cloud service provider infrastructure as well as interfaces between internal systems and clouds.

Governance—Focuses on prevention, detection, and mitigation policies. Threat intelligence, for example, can monitor and prioritize threats for SMBs and businesses to ensure critical systems are fully protected. On the other hand, individual cloud customers can benefit from security policies and training for user behavior. They are mainly used in organizational settings, but safe use and threat response guidelines can be beneficial to all users.

A cloud-Native Breach is a data breach in the cloud. Unlike the initial failure of this stolen data, it is often generated using the functionality of the default cloud. Cloud violations have exploited errors or vulnerabilities in cloud deployments without malware, and "expanded" to "expand" location data is "exfiltrate."

Misconfiguration—Elegant violations are often on customer cloud responsibility for security, including cloud service configuration. According to the study, only 26% of the current IaaS environment can now identify the current IaaS environment Shows. Regardless of the power of IaaS, it often played a role in the front door for a cloudy violation and the attacker will be able to land successfully and continue to expand the data expansion and export. Research also shows 99% of incorrect compounds and keeps noticeable in the cloud customer's IaaS.

A data recovery—cybersecurity plan is required to protect the effectiveness of critical negative disorders. Disaster recovery plans include policies, procedures, and tools to ensure data recovery and organizations' ability to continue operations and businesses.

Internal Threat—Injection You can expose your organization to violate cyber security using cloud services. A recent risk report found that 85% of organizations have an irregular activity that indicates insider threats.

1.5.4 Cloud Security

Traditional IT security has undergone significant changes due to the migration to cloud computing. The cloud paradigm is more convenient, but the always connection requires additional security issues. Cloud security is different from the old IT approach in several ways as a modern cybersecurity solution.

Data storage: The basic difference is that previous IT architectures depended primarily on the storage of local data. Organizations have long recognized that it takes a long time and is costly to develop their own IT infrastructure for accurate and configurable security measures. Cloud-based frameworks have helped charge system development and maintenance fees, as well as remove user control.

Scaling Speed: In a similar way, when the IT system is scaled, cloud security must be a unique week. The central infrastructure and application are very modular and fast for mobilization. This capability can be uniformly customized due to organizational changes, but the ability of the organization to occur if the organization is required for updates, and the ability to maintain safety and maintain safety.

Figure 1.3 shows the different tools for Cloud Data Protection. There are tools such as Data Discovery, Asset and Data Classification, Encryption, Key Management, Data Discovery, User Analytics, Access Control, Authentication, Configuration Hardening, Logging, and Reporting which are used for cloud security.

End-User Interfacing System: The same cloud system has interfaces with many different systems and services that must be safeguarded for companies and individual users. Access privileges must be managed from the ENGUSER device level to the system level to the network level. Moreover, providers and consumers should be cautious when it comes to vulnerability. This can be achieved by using dangerous configurations and access to the system.

Proximity to other network systems and data: Because cloud systems are always interconnected between cloud services and all consumers, the appropriate network may be built by the supplier. You may infect the remainder of the network landscape by infecting one vulnerable device or component. Suppliers are relieved of additional network security responsibilities. These additional items give products that only exist within a long-term system rather than their own.

The majority of cloud security vulnerabilities are caused by users and cloud suppliers in the private and company context. In cyber security, both must act as casters. This triple technique necessitates the interpretation of the customer and the supplier.

Fig. 1.3 Tools for Cloud Data Protection

Configuration and maintenance of security systems. User security education is behavior and technicality. Cloud and user providers must be transparent and responsible for ensuring the safe maintenance of both parties.

1.5.5 Attacks on the Cloud

1.5.5.1 Cloud Computing Security Vulnerabilities

The corporation swiftly leverages clouds for the Transformation of Distribution Directly.

There are certain benefits to cloud computing, such as faster processing times in public clouds, which can boost productivity when contrasted to more limited options in cloud storage (more limited access in dedicated clouds). However, despite the benefits of the cloud solution, data protection is one of the major challenges that prevent the organization from using cloud solutions. According to the results of this poll, 93% of firms are fairly susceptible to cloud computing hazards and are unduly concerned. The cloud infrastructure may be tough, and we all know that complexity is the adversary of security.

Most cloud security experts believe that businesses may benefit from the cloud's built-in security solutions, but organizations can make major mistakes that expose crucial data and systems. Some of the most typical cloud security problems are utilizing solder accounts inappropriately, inner system weaknesses, and employee permissions. Unauthorized APIs and dangerous APIs are connected to the fact that this one is the lone cloud security problem (42% of respondents).

(1) **Account Hijacking**: Account hijacking is a cloud threat that takes the user's credentials. It is also known as session abduction. There are several assailants.
(2) **Phishing**: Do not instruct the user to a non-secure website, steal the information, or spray the session identifier.
(3) **Key recorders**: Record user-created keystrokes, including user IDs and passwords, and send attack information.
(4) **Buffer Overflow Attack**: Overwrite memory data with malicious data in order to exploit unauthorized access attackers.
(5) **Cross-site scripting (XSS) Attack**: An injection attack in which attackers deliver harmful situations to a susceptible account via a web browser.
(6) **Brute force attack**: an attacker gains access to an account by guessing the password—typically using the Software.
(7) **Data breaches**

During the first half of 2019, there were at least 3,800 data infractions. These breaches resulted in approximately 4.1 billion exposed records, a 54% increase over the past year. Data breaches are an issue that may cost millions of dollars every time they occur. According to Verizon's 2019 Data Breach Investigation Report, small companies accounted for 43% of victims. One of the primary causes for data breaches in small organizations is that they would not have the same level of security as large corporations. They are easy prey and bear the brunt of the consequences when information is breached.

The result of data failures can include:

- Negative impact on partner, customer, and customer brand and trust loss.
- Important intellectual property rights are lost.
- Terminal discovery and other services.
- Legal measures.

(8) **Insecure API**

The Custom Application Interface (API) is a popular method of optimizing cloud computing. In general, the APIs used in the office may be utilized to quickly transfer information across two or more programs. It is well-known for its ease of use and ability to boost productivity, and the API might be the root of the cloud vulnerability. An attacker may readily access corporate data and launch DDoS assaults by using dangerous APIs. When beginning API assaults, more sophisticated attackers might utilize various techniques to prevent detection. As organizations' reliance on APIs grows, so does the number of assaults.

(9) **Malicious Insider**

Even if you guard against various sorts of cloud security assaults, you may still be subject to malicious insider threats, both current and prior.

- A worker
- Builders and contractors
- Business Associates

You can steal information, ruin IT systems, and meddle with IT systems if you have illegal access to the system. Insider attacks have grown by 47% during 2018, according to the Ponemon Report 2020. Inner attack costs have risen by 31% since 2018. Zlons made up just 23% of the insider threats. The majority of insider assaults are the result of human mistakes.

(10) **System vulnerability**

System vulnerabilities can occur for more common cloud security vulnerabilities. The integration of unauthorized applications in the third shell may cause the system's risk, or the security tools for the cloud system are not well. The common system can be negatively affected by the cloud service:

- User input lacks input confirmation.
- Inadequate registration and monitoring
- Incorrect error handling
- Do not disconnect the database connection.

(11) **Contract violation with customers**

Cloud computing contracts are cunning. This is often available, so you limit the restrictions on whether you have permission to access data, where you can access it. If the employee moves the data to the cloud without permission, it may violate the business agreement and violate the legal action. For example, if a customer service provider supports the right to share all data in the cloud of a third party, it violates the company and the confidentiality agreement. This causes data to be leaked by customers, employees, and other stakeholders that can be loaded into the cloud.

(12) **Services Outages**

The cloud infrastructure is a wide range, but a failure is a high cloud. Hardware problems and configuration supervisors cause this termination. The same problem to collecting traditional local data centers.

The cloud can be attacked by distributed denial services and other malicious mechanisms designed to deteriorate the presence of cloud resources and services. If an attacker cannot use public cloud resources or services, it affects each business or cloud user that uses these resources and services.

Cloud providers can be adapted when processing attacks and support groups may be helpful when a specific business workload is attacked. Other users in companies and public clouds cannot prevent cloud shutdowns and attacks but consider the impact on cloud and data source workloads and plan these activities within the framework of disaster recovery strategies.

Given the extensive nature of public clouds, you can resolve emergency recovery through high availability architecture implemented in clouds and areas. However, such posture is not automatic, it should be not dependent on the development and depend on methods.

(13) **Weak encryption**

Cloud companies use encryption techniques to protect data stored on their servers, but they generate random encryption data using a limited randomness provider (such as time). Linux-based virtual machines produce arbitrary keys in milliseconds. However, this may not be enough to securely encrypt data because attackers use complex decoding algorithms to damage information. Therefore, cloud computing developers need to consider how to protect their data before sending it to the cloud.

1.5.6 Attack Vectors for Cloud Computing

The main purpose of cyberattacks in cloud computing is to access customer data while preventing access to cloud services. Both can significantly harm cloud users who rely on cloud service security. When a cloud company is attacked, the hacker often reverses interaction between cloud users and services or apps.

- Vulnerability operation of cloud computing.
- Stole user credentials somewhere outside the clouds.
- After hacking a user's password, use a preliminary legal approach to the cloud.
- Play a malicious role.

Cloud computing services can be attacked in a variety of ways, and hackers are constantly pushing for more complex development. However, even the most basic understanding helps the cloud developer design a safe solution. Below is a list of 10 types of cyberattacks that are most commonly committed to cloud users.

(1) **Cloud malware injection attack**

Malicious injection attacks have complete control over cloud user data. To do this, hackers insert the affected module of the implementation of the service into the SaaS or PaaS system or insert a virtual machine instance into the IaaS solution. If the cloud system has been successfully denied, lead the cloud user to a component or a hacker instance where the malicious code will run. The opponent can then initiate harmful actions such as changing data or tapping. A scenario assault of Cross-site and SQL Injection Attack is the most typical injection attack of malicious applications. Cross-site script attacks include hackers inserting malicious programs (Flash, JavaScript, etc.) into susceptible websites. In 2011, German researchers coordinated an XSS assault on Amazon Web Services' cloud computing architecture. In the case of SQL injection, an attacker targets a SQL server with a weak database application. In 2008, the Sony PlayStation site was a victim of a SQL injection attack.

(2) **Cloud Service Abuse**

Hackers can use low-cost cloud services to attack specific individuals, businesses, and cloud providers. In 2010, for example, Brian and Anderson security specialists carried out a DoS attack using Amazon EC2 cloud architecture. As a result, by paying only $6 in renting a virtual service, they could provide their clients inaccessible on the Internet. Thomas Roth performed a brute-force assault at the 2011 Black Hat Technical Security Conference. Hackers can use strong cloud resources to provide millions of credentials to target user accounts by renting cloud company servers.

(3) **Denial of service attack**

DoS attacks are designed to overwhelm infrastructure and disable user services. These attacks are particularly dangerous for cloud computing platforms. This is because even a single cloud server can overwhelm and harm many users. As workload increases, the cloud system deploys new virtual servers and service instances and begins to provide additional computing capacity. Cloud systems are trying to stop cyberattacks. Finally, cloud platforms will slow down, preventing real users from accessing cloud services. If hackers deploy more zombie computers to target multiple systems, DDoS attacks in cloud settings can be even more harmful.

(4) **Side-channel attack**

When a hacker puts a hostile virtual machine on the same server as the targeted virtual machine, a side-channel assault is prepared. Hackers target cryptography algorithm system applications during side-channel attacks. However, a secure system design can prevent such dangers.

(5) **Wrapping attack**

In a cloud setting, a wrapping assault is a form of a man-in-the-middle attack. Because consumers often access services through web browsers, cloud computing is prone to wrapping attacks. XML signatures are used to protect user credentials from illegal access, but they don't protect the location of the document. As a result, an attacker can modify an XML document by utilizing the XML signature element. For instance, in 2009, a flaw in Amazon ELASTIC COMPUTING (EC2) was discovered. This weakness was able to change the messages sent by the successful packaging attack of an attacker.

(6) **Man-in-the-cloud attack**

In this type of attack, the hacker accesses and manipulates cloud services, exploits flaws inside the synchronization token system, and the synchronization token is a unique configuration that allows an attacker access. Users may never know if their account has been hacked because attackers can get authentic synchronization tokens at any time. Furthermore, hijacked credentials may never be retrieved.

(7) **Insider attack**

Insider attacks are carried out by real users who deliberately violate security restrictions. In a cloud system, an attacker can be a cloud provider administrator or an employee of the client company with significant rights. To avoid this malicious behavior, cloud developers need to design their security infrastructure with multiple access levels to cloud services.

(8) **Accounts or services hijacking**

The hijacking of accounts or services is performed after accessing users. There are many ways to achieve this and snatch spyware fishing for cookies addiction. Once the cloud account is hacked, an attacker can receive personal information about user or company data and damage cloud computing. For example, a SaaS provider, Salesforce, was a victim of phishing fraud which led to the impact of the company by all customers in 2007.

(9) **Advanced persistent threats (APTs)**

APTS is a theft attack in which hackers take private data from a cloud or cloud service that has no authorized users. If such assaults continue, hackers can modify the security systems in place to combat them. Once illegal access is gained, hackers can utilize network traffic to conduct destructive actions via data center networks.

(10) **New attacks: Specter and Meltdown**

These two forms of assaults first surfaced early this year but have quickly become a new danger to cloud computing. By using a design flaw in most current CPUs, an attacker can use malicious Scripts to extract encrypted data from memory. Specter and Meltdown, both compromise the separation of application programs so that attackers can retrieve kernel data. This is a major concern for cloud developers because not all cloud users apply the latest security updates. Specter and Meltdown both break the barrier between programs and the system software, allowing attackers to access data from the core. This is an important issue for cloud storage developers since not all cloud platform customers apply the latest security fixes.

1.6 Applying Machine Learning Techniques for Cloud Security

AI and machine learning [14] cannot be a bullet, but you can play a key role in the cloud security strategy. The speed of adopting clouds is now moved only in most organizations with cloud processes within the work. However, as the threat of cybersecurity becomes increasingly complex, cloud workloads must be secure because more investment is needed. Most of these investments are available for the possibility of AI [15] and machine learning (ML). Although no one can completely reduce violations, AI and ML are provided by Google and AI programs, although

AI and ML can improve overall IT operations protection and detection of real-time data analysis and threat threats. Here are some examples of how these technologies can help your cybersecurity strategy.

(1) **Big Data Processing**

Cybersecurity systems [16] generate more data than human teams can study and analyze. Machine learning technologies detect threats using all this data. The more data you process, the more patterns it detects, remembers, and uses to detect normal pattern flow changes. Such changes may become cyber threats. A training machine should be noted, for example, that regular and employees are regularly typed and considered normal from getting into the system and entering other traffic and user behavior models. It is respected to escape these norms, such as the entrances of the first morning wandering. This means the potential threat can be assigned and considered faster.

(2) **Event Prediction**

When using more Data-Driven approaches, artificial intelligence is used to detect and actively operate warnings and actively operate or use them in the future. This is a work based on the known behavior and known actions based on known actions, the acknowledgment of threats based on the known threat, and the analysis of data from protected endpoints. This allows more intelligent approaches to collect all data as well as "bad" activities, to enrich the root cause of potential attacks in other sources and minimize effects immediately when attacks are detected. It can also help you create a short cycle between detection and modification by ensuring the ability of the team's security to respond faster with the best data.

(3) **Event Detection and Blocking**

If an AI and technological learning system that processes data collected through the system is acquired through the system, you can warn or respond to individuals by turning off a particular user's user, among other alternatives. These measures are carried out, and occurrences are often identified and prevented for several hours to avoid the potentially harmful flow of code into the network and data leakage. Companies can respond to security events using this technique, which has been analyzed and associated with real-time geographic data, allowing the company to possibly get warning work.

(4) **Automatic Technical Delegation**

Warnings for potential threats or ideals have enormous potential to remove many noises that can be quite common to many security platforms but are quite common to important technologies. The security team focuses on more important and complex threats for AI and processing systems for routine tasks and first Level security analysis. This is especially important when considering the lack of current technology of cyber security. With 51% of organizations that approve the problems of Cybersecurity technology, companies will allow enterprises to commission the first level of the archive so that the security experts can try to focus their efforts to fight complicated attacks. This does not mean that these technologies can replace personal analysts. Therefore, cyber-attacks often occur in human and machinery efforts, so they need answers to people and cars. However, the analyst can specify the priority of the workload and obtain the work more efficiently.

1.6.1 Types of ML Algorithms to Detect Attacks and Malicious Behavior

Machine Learning algorithms have been known to improve security in cloud computing services [17]. To use ML models for prediction, it's important to analyze available data in real-time, identify threats and network attacks, and generate labels using the supervised learning method. The below paragraph shows various supervised learning algorithms [18] that have been used for providing good results in improving cloud security issues.

1.6.1.1 Supervised Learning Algorithms

Supervised Learning algorithms work on the labeled data and provide results based on the target variables.

(1) K-Nearest Neighbor

K-Nearest Neighbor (KNN) is a classification algorithm that is mainly used to classify data based on the information provided and based on the classification the users can make assumptions about the data and provide a solution to the problem. KNN uses a similarity measure such as distance-based calculation to classify or group data based on the similarity score. Ensuring cloud security is always a top priority and remains a challenge for many enterprises. Various frameworks have been used to enhance the cloud security [19] in the given data, such as the data encryption method. However, these frameworks do not always provide better security due to the data being confidential. Zardari et al. [20] showed a classification-based approach keeping in mind the confidentiality of the data. The main objective of this document was to protect and protect the information collected from the data. To achieve this goal, the KNN algorithm was used to group information based on cloud management and virtual security issues. Grouping was done using two classes, one class containing sensitive information and the other containing non-sensitive information. This ensured that the sensitive data is maintained and there is no loss of information. Using the KNN technique as a classifier aided in the achievement of safe information by maintaining the order of the information in the computer environment.

Sensitive data necessitates greater security than less sensitive information; hence, KNN gave a superior solution in the cloud computing system while keeping data privacy and confidentiality in mind.

Another sector that uses ML algorithms for security threats and attacks in CC is the Intrusion detection and prevention system (IDPS). Earlier cybersecurity systems were used to prevent security threats and attacks on the cloud network [21] from cyberbullies. However, using the previous methods in IDPS is not adequate to identify and manage security threats. Luckily AI and ML provide a good solution to these types of problems. Algorithms like ANN, KNN, and GA have been shown to provide effective results in cyber security [22].

Cyberbullies have been very proficient in designing new systems to cover up for the malware attacks being sent through the systems which often went misdiagnosed by the algorithms. Hence an improved system was needed to detect such kinds of attacks and prevent them from disrupting the security and privacy of the data [23]. Healthcare system data is highly confidential as all the information is being stored in the cloud. The author [24] demonstrates the use of the KNN algorithm to identify k data patterns having the highest similarity based on privacy and security issues of the cloud data in the healthcare system such that the model can maintain the security level and prevent any threats to the cloud data.

(2) Artificial Neural Network (ANN)

ANN in cloud security [25] is used as a supervised approach to analyze the computing environment and recreate how humans think in terms of accessing information from the data provided. In a nutshell, ANN works like a human brain in terms of solving difficult challenges which would otherwise be impossible for a human to detect.

Elzamly et al. [26] recommended the usage of ANN for detecting security-related vulnerabilities in the banking sector's cloud computing environment. Because ANN gains new skills throughout the training phase, its neural network capabilities make it useful in security-related difficulties. The representation of the level of cloud security was predicted using Levenberg–Marquardt (LMBP) process. LMBP is a nonlinear representation of the model which is used to predict the level of security present in the cloud environment. This is done by measuring the correctness of forecasted results and reducing the error rate between the actual and predicted results using mean squared error (MSE). The model used for extracting useful information was ANN and LMBP was used as a measure to evaluate the cloud security [27] and predict the results based on the information provided. LMBP algorithms have been proved to be incredibly efficient for testing and planning methods in the banking sector related to CC security.

Because of the composite and distributed architecture of cloud frameworks, it is difficult to detect complex threats in cloud environments. Moreover, a variety of portable registration and storing devices are linked to the cloud architectures to support a client's requirements, which in turn adds to the complexity and difficulty of identifying digital attacks. Sayantan et al. [28] are capable of identifying major types of digital attacks in the cloud environment [29] which affect the security of cloud computing. The author proposed a framework that uses ANN to reduce the attacks significantly by analyzing the traffic information about the cloud environment and stopping the system traffic which contains unauthenticated information to pass through the cloud environment. In this paper [30] the author uses both supervised and unsupervised ANN capabilities on encrypted data extracted from a semantically secured cryptosystem based on the characteristics of homomorphic encryptions. The lack of interest in the information from the owners is the main cause of the instability of data from the outsiders who are responsible for storing and managing the cloud data in the computing environment. Employing ANN in various fields has helped to achieve considerable security and prevented the systems from major threats and attacks in the cloud system.

(3) Support Vector Machines (SVM)

SVM is the most common supervised learning algorithm, popularly used for classification and regression problems. The main aim of SVM is to separate the data into classes by using a hyperplane to divide the data into two classes. The output of SVM will contain the data separated into two classes with an expectation to be reasonably far based on the distribution technique. Recently many studies have demonstrated the use of SVM for cloud security and services. Grusho et al. [31] identified major security attacks on cloud-computing systems using ML models. Identifying such malicious unwanted intruders in the cloud environment is important as it would disrupt the safe use of could services for the users. Nowadays, due to the high use of virtualization in the world for different businesses and other sectors, customers expect utmost security and trust in the system to provide a continued undisrupted environment of cloud computing. Various solutions have been proposed to capture data leaks and breaches, unknown risk of profiles being created, DDoS attacks on the system when transferring to a cloud environment. The proposed solution is used to identify such attacks and accumulate the incident details and feed them into the model to intercept them and prevent them from future attacks. IDPs systems have been widely used to identify such problems and detect such unwanted threats. This type of information when fed to the model such as SVM demonstrated a highly robust system and is extremely useful in solving security-related threats and potential risks in the environment. In a recent study [32], the author used different ML techniques including SVM, Naïve Bayes, Random Forest (RF) as a classification algorithm to detect unwanted attacks and intruders in the system. Using different ML algorithms and comparing the results, demonstrated that SVM provides the best results with an overall accuracy of around 99% in classification and detection of intrusions based on the dataset used.

(4) Naïve Bayes

The naïve Bayes classifier uses Bayes' theorem to identify irrelevant features from a set of relevant features to demonstrate the unrelatedness of the feature in the data. Naïve Bayes has been used in many sectors of the world, however, for the cloud security services, Naïve Bayes has been used in the detection of DDoS attacks and threats in many studies.

To combat DDoS threats, Zekri et al. [33] developed a distributed DDoS detection system using the C4.5 algorithm. Hidden bugs and vulnerabilities can be exploited by attackers that exist in underlying inventions and creations of the system. Such attacks can be harmful to the cloud environment and can cause serious damage to cloud security. ML algorithms such as Naïve Bayes have been used to identify such attacks and prevent them from causing any destruction to the working of the cloud environment. Traditional approaches such as Intrusion detection system (IDS) is not capable to perform well, given such an adverse situation and ever-changing environment. Hence, a more robust algorithm such as Naïve Bayes outperforms the traditional approaches used to prevent security attacks in the cloud.

Hanna et al. [34] examined and described strategies to alleviate CC security concerns as a first step toward achieving a secure and safe CC environment. This main goal was to spread awareness and demonstrate the importance of developing a secure cloud environment and establish necessary steps to be taken towards achieving a secured distributed computing environment. The results showed that utilizing a decision tree model [35] establishes a powerful strategy that allows the leader to access the degree of the cloud while ensuring that different types of assistance will be provided based on the type of security issue identified by the model. Supervised learning algorithms like naïve Bayes, decision trees, multilayer perceptron, SVM, and partial tree (PART) have been shown to provide high accuracy and performance while maintaining security issues in cloud computing. In [36], the author proposed a combination method based on anomaly detection techniques to identify and detect attacks in mobile computing. Naïve Bayes algorithm was used as the final algorithm as it showed improvements and provided efficiency based on other supervised learning methods.

1.6.1.2 Unsupervised Learning Algorithms

In unsupervised machine learning, the models are trained without any label or supervision just like in supervised learning. More broadly put, unsupervised ML is a grouping evaluation that utilizes data for examining the results without requiring any label to be defined in advance. The most common uses of unsupervised algorithms are clustering and dimensionality reduction.

(1) K-means Clustering

There are various clustering methods available such as density-based, hierarchical clustering, etc., but out of those K-means is the most widely used and straightforward unsupervised technique. The main aim of k-means is to find k groups in a dataset, where k is a fixed number assigned to the model. The major challenge faced by many researchers is to select an appropriate value of k for the given dataset. A group here is defined as the accumulation of information based on certain similarities being observed in the dataset. Grouping these similarities provides an important meaning to the dataset. Recent studies have used k-means clustering to overcome cloud security in the environment. In this study [37], the author focused on the issues related to the attacks in the cloud environment and provides an analysis on how to disrupt the attacks and overcome the security-related threat. K-means algorithm was used to group those attacks and provide improved security to the cloud. Ahmad et al. [30] conducted a survey on machine learning algorithms for demonstrating the achievement of cloud security using ML. Due to data vulnerability by third parties who store, handle, and analyze the data, the owner's interest diminishes as they do not have much control over the data. Due to this, the possibility of abuse still exists. Secure multi-party communication was used to process the data securely as the data owners in the cloud require data privacy as well as efficiency in cloud management.

(2) Singular Value Decomposition (SVD)

SVD is one of the most common types of dimensionality reduction technique followed by Principal Component Analysis (PCA). SVD is a matrix factorization technique that uses grid values to be transformed into a vector. SVD is one of the most powerful tools that exist today and can be used in almost any field. It provides an answer to a meaningful problem that can fit into a model and give accurate results. Kumar et al. [38] discussed the methodology, problems, and unanswered concerns around the successful operation of machine learning-based security detection in the cloud. Investigators looking into security incidents in the cloud would not know what normal and abnormal irregularities in security are. Hence, relying on them would not produce satisfactory results. Even though ML models have shown improvements, due to the lack of benchmark datasets, model evaluation still raises certain concerns. It is necessary to use other algorithms along with SVD to provide enough features for identifying security issues in cloud computing. This provides more clarity and efficiency in delivering more accurate results when looking into security threats in CC. Feng et al. [39] proposed the use of SVD methods for dimensionality reduction in cyber security data which usually contains high dimensional information. To provide efficiency it is useful to reduce the dimensions using secured orthogonal tensor SVD and then outsource the task to the cloud to provide high security to the users.

1.7 Conclusion

In this study, we provided a systematic review of cloud computing techniques and security issues faced in the cloud environment. The review analyzed various aspects of cloud computing and focused on discussing security threats and attacks faced by the cloud environment. Different types of supervised and unsupervised ML algorithms were discussed to overcome the security threats and identify malicious attacks on the cloud server. Based on the recent studies, we can conclude that the models perform well on a particular type of dataset to solve security challenges and identify different threats and attacks. However, more research is needed which involves different types of datasets that include thorough feature selection/feature extraction methods. Hence, we conclude that both supervised and unsupervised methods have been shown improvements in the detection and prediction of the model's performance. However, in the future, we would like to perform a detailed study using unsupervised learning models like clustering, anomaly detection, and neural network to enhance the model's performance and expand our research on cloud security.

References

1. A. Gordon, The hybrid cloud security professional. IEEE Cloud Comput. **3**(1), 82–86 (2016). https://doi.org/10.1109/MCC.2016.21
2. A. Qayyum et al., Securing machine learning in the cloud: a systematic review of cloud machine learning security. Front. Big Data **3**(November) (2020). https://doi.org/10.3389/fdata.2020.587139
3. M. Amar, M. Lemoudden, B. El Ouahidi, Log file's centralization to improve cloud security, in *2016 2nd International Conference on Cloud Computing Technologies and Applications (CloudTech)* (2016), pp. 178–183. https://doi.org/10.1109/CloudTech.2016.7847696
4. J. Soni, S.K. Peddoju, N. Prabakar, H. Upadhyay, Comparative analysis of LSTM, one-class SVM, and PCA to monitor real-time malware threats using system call sequences and virtual machine introspection, in *International Conference on Communication, Computing and Electronics Systems. Lecture Notes in Electrical Engineering*, vol 733, ed. by V. Bindhu, J.M.R.S. Tavares, A.A. Boulogeorgos, C. Vuppalapati (Springer, Singapore, 2021), pp. 113–127. https://doi.org/10.1007/978-981-33-4909-4
5. P. Gangwani, J. Soni, H. Upadhyay, S. Joshi, A deep learning approach for modeling of geothermal energy prediction. Int. J. Comput. Sci. Inf. Secur. (IJCSIS) **18**(1) (2020)
6. L. Auria, R.A. Moro, Support Vector Machines (SVM) as a technique for solvency analysis. SSRN Electron. J. (2008). https://doi.org/10.2139/ssrn.1424949
7. G. Stein, B. Chen, A.S. Wu, K.A. Hua, Decision tree classifier for network intrusion detection with GA-based feature selection, in *Proceedings of the 43rd Annual Southeast Regional Conference on—ACM-SE 43* (2005), vol. 2, p. 136. https://doi.org/10.1145/1167253.1167288
8. T. Bhardwaj, H. Upadhyay, L. Lagos, deep learning-based cyber security solutions for smart-city: application and review, in *Artificial Intelligence in Industrial Applications*, vol. 25, ed. by T. Sharma, S. Fernandes (Springer, Cham, 2022)
9. S. Muthurajkumar, S. Ganapathy, M. Vijayalakshmi, A. Kannan, Secured temporal log management techniques for cloud. Procedia Comput. Sci. **46**, 589–595 (2015). https://doi.org/10.1016/j.procs.2015.02.098
10. S. Muthurajkumar, M. Vijayalakshmi, S. Ganapathy, A. Kannan, Agent based intelligent approach for the malware detection for infected cloud data storage files, in *2015 Seventh International Conference on Advanced Computing (ICoAC)* (2015), pp. 1–5. https://doi.org/10.1109/ICoAC.2015.7562810
11. B. Jin, Y. Wang, Z. Liu, J. Xue, A trust model based on cloud model and Bayesian networks. Procedia Environ. Sci. **11**, 452–459 (2011). https://doi.org/10.1016/j.proenv.2011.12.072
12. N.S. Selamat, F.H.M. Ali, Comparison of malware detection techniques using machine learning algorithm. Indones. J. Electr. Eng. Comput. Sci. **16**(1), 435–440 (2019). https://doi.org/10.11591/ijeecs.v16.i1.pp435-440
13. G. Ramachandra, M. Iftikhar, F.A. Khan, A comprehensive survey on security in cloud computing. Procedia Comput. Sci. **110**, 465–472 (2017). https://doi.org/10.1016/j.procs.2017.06.124
14. D. Gangwani, P. Gangwani, Applications of machine learning and artificial intelligence in intelligent transportation system: a review, in *Lecture Notes in Electrical Engineering*, Springer (2021), pp. 203–216
15. T. Bhardwaj, R. Mittal, H. Upadhyay, L. Lagos, Applications of swarm intelligent and deep learning algorithms for image-based cancer recognition, in *Artificial Intelligence in Healthcare* (Springer, Singapore, 2022), pp. 133150
16. P. Gangwani, A. Perez-Pons, T. Bhardwaj, H. Upadhyay, S. Joshi, L. Lagos, Securing environmental IoT data using masked authentication messaging protocol in a DAG-based blockchain: IOTA tangle. Futur. Internet **13**(12), 312 (2021). https://doi.org/10.3390/fi13120312
17. T. Bhardwaj, C. Reyes, H. Upadhyay, S.C. Sharma, L. Lagos, Cloudlet-enabled wireless body area networks (WBANs): a systematic review, architecture, and research directions for QoS improvement. Int. J. Syst. Assur. Eng. Manag. (2021). https://doi.org/10.1007/s13198-021-01508-x

18. D. Gangwani, Q. Liang, S. Wang, X. Zhu, An empirical study of deep learning frameworks for melanoma cancer detection using transfer learning and data augmentation, in *2021 IEEE International Conference on Big Knowledge (ICBK)* (2021), pp. 38–45. https://doi.org/10.1109/ICKG52313.2021.00015
19. T. Bhardwaj, H. Upadhyay, S.C. Sharma, Framework for quality ranking of components in cloud computing: regressive rank, in *2020 10th International Conference on Cloud Computing, Data Science & Engineering (Confluence)* (2020), pp. 598–604. https://doi.org/10.1109/Confluence47617.2020.9058016
20. M.A. Zardari, L.T. Jung, N. Zakaria, K-NN classifier for data confidentiality in cloud computing, in *2014 International Conference on Computer and Information Sciences (ICCOINS)* (2014), pp. 1–6. https://doi.org/10.1109/ICCOINS.2014.6868432
21. T. Bhardwaj, H. Upadhyay, S.C. Sharma, Autonomic resource provisioning framework for service-based cloud applications: a queuing-model based approach, in *2020 10th International Conference on Cloud Computing, Data Science & Engineering (Confluence)* (2020), pp. 605–610. https://doi.org/10.1109/Confluence47617.2020.9058266
22. D. Richards, The Benefits of Artificial Intelligence on Workplace Productivity. *Mavinlink* (2017)
23. S. Shamshirband, A.T. Chronopoulos, A new malware detection system using a high performance-ELM method, in *Proceedings of the 23rd International Database Applications & Engineering Symposium on IDEAS'19* (2019), pp. 1–10. https://doi.org/10.1145/3331076.3331119
24. J. Park, D.H. Lee, Privacy preserving k-nearest neighbor for medical diagnosis in e-health cloud. J. Healthc. Eng. **2018**, 1–11 (2018). https://doi.org/10.1155/2018/4073103
25. T. Bhardwaj, H. Upadhyay, S.C. Sharma, An autonomic resource allocation framework for service-based cloud applications: a proactive approach, in *Soft Computing: Theories and Applications* (2020), pp. 1045–1058
26. A. Elzamly, B. Hussin, S. Abu Naser, K. Khanfar, M. Doheir, A. Selamat, A. Rashed, A new conceptual framework modelling for cloud computing risk management in banking organizations. Int. J. Grid Distrib. Comput. **9**, 137–154 (2016). https://doi.org/10.14257/ijgdc.2016.9.9.13
27. T. Bhardwaj, H. Upadhyay, S.C. Sharma, Autonomic resource allocation mechanism for service-based cloud applications, in *2019 International Conference on Computing, Communication, and Intelligent Systems (ICCCIS)* (2019), pp. 183–187. https://doi.org/10.1109/ICCCIS48478.2019.8974515
28. S. Guha, S.S. Yau, A.B. Buduru, Attack detection in cloud infrastructures using artificial neural network with genetic feature selection, in *2016 IEEE 14th International Conference on Dependable, Autonomic and Secure Computing* (2016), pp. 414–419. https://doi.org/10.1109/DASC-PICom-DataCom-CyberSciTec.2016.32
29. T. Bhardwaj, S.C. Sharma, An autonomic resource provisioning framework for efficient data collection in cloudlet-enabled wireless body area networks: a fuzzy-based proactive approach. Soft Comput. **23**(20), 10361–10383 (2019). https://doi.org/10.1007/s00500-018-3587-x
30. A.N. Khan, M.Yu. Fan, A. Malik, R.A. Memon, Learning from privacy preserved encrypted data on cloud through supervised and unsupervised machine learning, in *2019 2nd International Conference on Computing, Mathematics and Engineering Technologies (iCoMET)* (2019), pp. 1–5. https://doi.org/10.1109/ICOMET.2019.8673524
31. A.A. Grusho, M.I. Zabezhailo, A.A. Zatsarinnyy, V.O. Piskovski, Secure automatic reconfiguration of cloudy computing. Syst. Means Inform. **26**(3), 83–92 (2016). https://doi.org/10.14357/08696527160306
32. V. Sharma, V. Verma, A. Sharma, Detection of DDoS attacks using machine learning in cloud computing. Commun. Comput. Inf. Sci. **1076**, 260–273 (2019). https://doi.org/10.1007/978-981-15-0111-1_24
33. M. Zekri, S. El Kafhali, N. Aboutabit, Y. Saadi, DDoS attack detection using machine learning techniques in cloud computing environments, in *2017 3rd International Conference of Cloud Computing Technologies and Applications (CloudTech)* (2017), pp. 1–7. https://doi.org/10.1109/CloudTech.2017.8284731

34. H.M. Said, I. El Emary, B.A. Alyoubi, A.A. Alyoubi, Application of intelligent data mining approach in securing the cloud computing. Int. J. Adv. Comput. Sci. Appl. **7**(9), 151–159 (2016). https://doi.org/10.14569/ijacsa.2016.070921
35. A. Mishra, N. Gupta, B.B. Gupta, *Security Threats and Recent Countermeasures in Cloud Computing* (2020), pp. 145–161
36. K. Arjunan, C.N. Modi, An enhanced intrusion detection framework for securing network layer of cloud computing, in *2017 ISEA Asia Security and Privacy (ISEASP)* (2017), pp. 1–10. https://doi.org/10.1109/ISEASP.2017.7976988
37. A. Meryem, D. Samira, E.O. Bouabid, Enhancing Cloud Security using advanced MapReduce k-means on log files, in *Proceedings of the 2018 International Conference on Software Engineering and Information Management—ICSIM2018* (2018), pp. 63–67. https://doi.org/10.1145/3178461.3178462
38. R.S.S. Kumar, A. Wicker, and M. Swann, Practical machine learning for cloud intrusion detection, in *Proceedings of the 10th ACM Workshop on Artificial Intelligence and Security* (2017), pp. 81–90. https://doi.org/10.1145/3128572.3140445
39. J. Feng, L.T. Yang, G. Dai, W. Wang, D. Zou, A secure high-order Lanczos-based orthogonal tensor SVD for big data reduction in cloud environment. IEEE Trans. Big Data **5**(3), 355–367 (2019). https://doi.org/10.1109/TBDATA.2018.2803841

Chapter 2
Energy Enhanced Routing Protocol for Heterogeneous Networks

Satyanarayana Mummana, Ramana Babu Budimure, Bosubabu Sambana, Preethi Gandreti, and A. Chandra Sekhar

Abstract The potential applications of wireless sensor networks (WSNs) in areas such as disaster management, battlefield monitoring, and border security surveillance have received more attention in recent years. Large numbers of sensor nodes, often left unattended, perform their tasks independently in such systems. Increasing the lifespan of a sensor network by decreasing its energy consumption, clustering is a crucial strategy. It can also improve the scalability of the network. Wireless sensor network researchers across disciplines prefer to assume that nodes are all the same, but in reality, a WSN's longevity and reliability may be improved by including nodes with varying amounts of available energy. In this study, we introduced the heterogeneous model for WSNs and discussed the clustering techniques that have been suggested for such networks (HWSNs).

Keywords Wireless sensor network · Cluster head · Homogeneous sensor network · Heterogeneous sensor network

S. Mummana · R. B. Budimure · B. Sambana (✉)
Department of Computer Science Engineering, Lendi Institute of Engineering and Technology (A), JNTU-GV, Vizianagaram, Andhra Pradesh, India
e-mail: bosukalam@gmail.com

P. Gandreti
GVP College for Degree and PG Courses (A), Engineering and Technology Program, Andhra University, Visakhapatnam, Andhra Pradesh, India

A. C. Sekhar
Avanthi Institute of Engineering and Technology, Tagarapuvalasa, Visakhapatnam, India

T. Bhardwaj et al. (eds.), *Artificial Intelligence in Cyber Security: Theories and Applications*, Intelligent Systems Reference Library 240,
https://doi.org/10.1007/978-3-031-28581-3_4

2.1 Introduction

A computer network, often known as a data network, is any communications network that facilitates the transfer of data between computers. Data is sent between connected computers in a network. Cable media or wireless media is used to establish the connections (network links) between nodes. The Internet is the most well-known system of its kind. Data is generated, transmitted, and ultimately received by nodes in a network. Hosts in PCs, mobile devices, servers, and other networking gear are all examples of nodes. When one device can communicate with another device, regardless of whether or not they are physically connected, we say that the devices are "networked." Connectivity to the Internet, file and print servers, facsimile machines, and electronic mail/instant messaging programs are all made possible via computer networks. Communications protocols to manage network traffic, size, topology, and organizational intent in the physical medium computer network utilized to convey their signal.

Because sensor nodes may be placed without any infrastructure and the network can monitor numerous dangerous or inaccessible regions that people cannot reach, wireless sensor networks (WSNs) have been employed in various fields in recent years. While routing, coverage, and localization have been the primary focus of recent WSN research [1], other areas have also been explored. Microsensor nodes that are cheap but yet have wireless capabilities and a low power need for data transmission are typical of WSNs, as are resource limits and battery life constraints. Given the low power reserves of sensor nodes, it is critical to use energy-saving strategies to prolong the usefulness of WSNs.

The network that resulted is said to be standardized. A pair of sensors would have the same life in these forms of networks if they had the same amount of energy [1, 2] used. However, some sensors use sensors with various capabilities and the resultant network is accordingly efficient results said to be heterogeneous. Homogenous sensors should not be taken for granted in the real world as sensing applications can require heterogeneous sensors and capabilities of connectivity to improve network efficiency and prolong the life of the network. Also, even though the sensors come with the same hardware, the same connectivity and sensing models cannot always be supported. In fact, at the production stage, two sensors that use the same platform are exactly the same there are no guaranteed physical properties.

With the capacity to deliver data anywhere, anywhere, and in every manner, wireless sensor networks have revolutionized the network environment. The Broadband Internet is just a minimal power supply that can be fitted to the sensor network node as an electronic unit. The Wireless Sensor Network (WSN) [1, 3] consists of a wide range of less-cost sensor nodes sensor sizes, local processing capacities, and wireless communication capabilities. The longevity of a sensor node is mainly determined by the final battery power supply of the original. Most of the WSN [3] protocols resume that sensors have the same storage, encoding, sensing, and sensing capabilities, and data communication. Three common [4] forms of heterogeneity occur, or

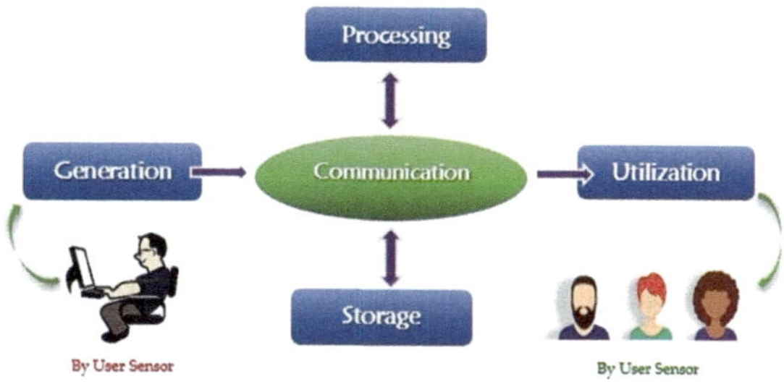

Fig. 2.1 Structure of sensor nodes

heterogeneity or energy heterogeneity is connected. The heterogeneity of the power implies that the node is driven by a line or battery that should be replaced.

Wireless sensor network routing varies in different ways from traditional fixed network routing. There is no infrastructure, there are cellular networks. Unlikely, sensor nodes and protocol for routing must comply with rigorous demands on energy conservation. In the Wireless Sensor Network Routing Protocol, the hierarchies of energy-efficient routing protocol are available to the WSN and Heterogeneous WSN categories (Fig. 2.1) [5].

2.2 Literature Survey

In recent years, wireless sensor networks have emerged as a promising new field of study. Together, the wireless sensor nodes in a sensor field and the sink make up a Wireless Sensor Network. Wireless sensor networks (WSNs) are made up of a large number of nodes that can collect data about their surroundings, do some basic calculations, and exchange data wirelessly. Detailed capabilities are outlined, including sensing, tracking, and alerting. By working together, these nodes can access it. These capabilities make wireless sensors ideal for tracking friendly forces in the field and keeping tabs on unusual occurrences, as well as regulating security and predicting traffic patterns. For these kinds of jobs, sensor networks must be extremely dependable. Recent years have seen a rise in the study of heterogeneous wireless sensor networks with the goal of making sensor networks more trustworthy [1]. The sensor nodes in a network may be made more scalable by forming clusters. The person in charge of a cluster is called the "cluster head" (CH). Both the sensors in a cluster and the network as a whole can vote on who will serve as the Cluster Head. The connection between clusters might be static or dynamic. For the sake of scalability and clear explanation, a variety of clustering algorithms have been developed specifically for Wireless Sensor Networks (WSNs). Ware-less sensor networks also make use of the concept of cluster-based routing to demonstrate efficient energy use (WSNs).

El Alami [6] recommended a secure election using three Fuzzy Parameters (SEFP) to optimize network existence and stability time. This method centered only on three fluctuating factors, such as the other control node. The discrepancy between the two

BS nodes and the total proximity. However, the sink distance and efficiency [7] are not critical parameters.

In the quest for the unified gridding hybrid for the upper-level chosen CH, Lee and Kao [8] suggested a semi-distributing clustering process. However, this approach does not have the confidence and credibility to select CH for data supply.

Saidu et al. [9] proposed an optimized cluster routing algorithm by considering the residual power of the network nodes while maintaining the maximum number of CHs over the whole network life. The network has been divided into an optimal number of CH clusters [7, 10]. Every CH selects the node with the highest remaining power in the cluster as vice CH. The CH turns the vice CH to sleep for the current cycle, and the vice CH becomes the CH for the next cycle. It considers, however, only the residual power of the nodes, while the other significant parameters were not taken into account in the CH election.

Haseeb et al. [11] proposed a Reliable Cluster-based Energy-aware Routing (RCER) protocol and Barman et al. [7] suggested a Secure Cluster-based Energy-aware Routing (SCER) protocol in order to improve the network life and reduce routing costs. Initially, the network was split into geographic growth clusters. The network is more effective in terms of electricity. RCER was then used as an optimum route to improve the next-hop election by accounting for the remaining space, the hop count and the weighted value of the Round Trip Time (RTT). In addition, routing instructions have been changed on the basis of the dimensions of the network. Support to ensure the reliability of the network. However, the length of the route and the average network throughput were smaller.

Khan et al. [12] proposed a Protocol that establishes a basis for the preference of CH for energy and transport multi-heterogeneity scenarios. It considers the initial power, residual energy, and traffic load of the nodes during the selection of CH, along with the average energy consumption of the network. The likelihood of being CH for a node during each round is formulated with a RE-TOPSIS traffic heterogeneity factor that depends on the parameters. This increases the life of the network and lowers the efficiency of the energy use network.

As a central calculation, LEACH has made a variety of improvements based on the different applications. A point-by-point description of LEACH and its predecessors occurs taking into account four important criteria, e.g. grouping strategy, aggregation of knowledge, and portability. Kindness and willingness to adapt. The LEACH convention randomly picks the cluster head and the sink does not know any details on the residual vitality of the device. Conveying remote technologies would deplete more strength as contrasted Gadgets built to get or sit down inactively.

The Broadband Internet has a minimal power supply that can fit the sensor network node as an electronic unit. The Wireless Sensor Network (WSN) consists of a wide range of less-cost sensor nodes sensor size, local processing capacities, and wireless communication capabilities. The longevity of a sensor node is mainly determined by a final battery power supply the original. Most WSN protocols presume that sensors have the same storage, encoding, sensing, and data communication capabilities. Three common forms of heterogeneity occur, heterogeneity or energy heterogeneity is connected. The power heterogeneity implies that a line or battery drives the node should be replaced. Wireless sensor network routing varies in different ways from traditional fixed network routing. There is no infrastructure; there are cellular networks. Unlikely, sensor nodes and protocols for routing must comply with rigorous demands on energy conservation.

Various techniques for deciding the practical cluster head have been considered from the above-mentioned writing that updates provide for implementation. However, critical metrics, such as leftover vitality, introductory vitality, and an optimal number of clusters in the method, have not been deemed an encouragement to CH determination, as we might reasonably know for alteration in the edge. Apart from being the primary and the other Convention, LEACH still has a few hindrances. The cluster head (CH) option is made haphazardly because a small node with higher and low vitality has an equal chance of being a CH. If a small node with poor energy is selected as CH, the dust can bite easily. To affect the heartiness of the method. In addition, the location and number of CHs are shifted in each round.

Khan et al. [12] have examined the Heterogeneity-aware Hierarchical Stable Election Protocol for WSNs (HSEP). HSEP is proposed to minimize energy transfer between the CH and the base station by reducing their gap. The distance reduction is necessary since, as the distance increases, the power of transmission increases. As the energy needed for transmitting is higher than the energy required for sensing and reducing transmission capacity, the network's life can be increased. HSEP has two forms of CHs: principal CHS and secondary CHs. Only primary CHs may engage in the collection of secondary CHs. The primary CHs are chosen based on the threshold; the advanced nodes have better chances. To be the main CHs. In the case of secondary CH collection, primary CHs search the distance between each other and submit data to only CHs at a minimum length; these minimum CH distances are called CHs. Data is transmitted from the nodes to the primary CHs that execute the aggregation operation and forwarded to the secondary CHs. Secondary CHs are sending the data to the base station.

Sharma and Sharma [13] proposed an EEE LEACH or Energy Efficient Extended LEACH is a multi-level clustering technique intended to improve energy efficiency by reducing its radio transmission wavelength. In this multilevel clustering strategy, besides providing a single layer of cluster formation between the nodes and the base station like LEACH, there are two layers of cluster formation. In the first Layer, CHs are created where regular nodes send their data to their respective CH, and CHs accumulate the obtained data using the fuse function. A great deal of interest has been paid to the science sector and its ability to serve a range of applications. However, due to sensor nodes' erratic behavior, cluster head-based data aggregation and routing problems face significant challenges for WSNs. Recently, many cluster head collection and routing strategies have been proposed to enhance network life.

Due to their ability to survive in a turbulent world without the requisite resources. The nodes in WSNs are deployed randomly in an ad-hoc way that helps collect sensing information from the outside world. These nodes observe, store, manage and forward the sensed data to neighboring nodes to be eventually accepted at the sink node. For effective data transfer to the base station, the whole network is split into small clusters, and this method is referred to as clustering [7]. After creating groups, the node that satisfies those criteria will be selected as the cluster leader.

Mummana and Rao [14] proposed a new model for data aggregation. The new technique reduces the distance to transmit the data from source to destination. The cluster head selection process was done using a metaheuristic optimization algorithm: Invasive weed optimization algorithm to find the best suitable and optimum cluster head, which reduces the data transmission distance and energy consumption and improves the network lifetime.

2.3 Related Works

Wireless network sensor routing varies in many ways from standard fixed network routing. There are no services, or internet networks unreliable sensor nodes can malfunction. The routing protocols must meet strict energy-saving requirements (Fig. 2.2) [13].

The hierarchies are in the Wireless Sensor Network Routing Protocol [6] in the Homogeneous WSN and Heterogeneous WSN classes.

Sensor nodes are grouped into clusters into a hierarchical or clustering network where the participants transmit data to the cluster heads. In contrast, the cluster heads act as relays to pass the data to the groups. LEACH Cluster-based method [9], CBR, LEACH, is a clustering-based protocol for EDG Low Energy Adaptive clustering hierarchy Wireless network data collection. Hundreds and thousands of wireless sensors capturing and distributing data are spread throughout the network. Cluster heads are often selected from the sensors to relay the gathered data to the base station in these networks. Each sensor node is cheap and easy and cannot be replaced by its low power level. This is why any sensor must take its turn to be an energy-efficient cluster head of the protocol.

A LEACH cycle is a round [9] with each round's phase set-up and continuous phase. The cluster head is created during the set-up process. Randomly, if the number is chosen to be less than any threshold T(n), in every sensor node in a range of between 0 and 1 is selected, the node is chosen to be the cluster leader. T(n) Formulae Following is:

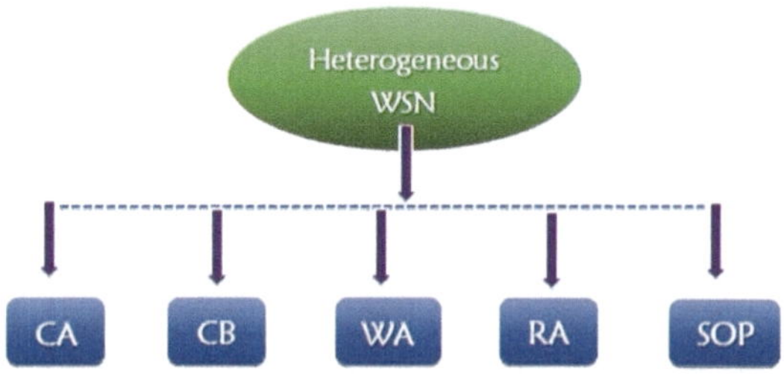

Fig. 2.2 Classification of various protocols in heterogeneous networks

$$T(n) = \begin{cases} \frac{p}{1-p\left(r \bmod \left(\frac{1}{p}\right)\right)} & \text{if } n \in G \\ 0 & \text{otherwise} \end{cases}$$

When p is the ratio of the cluster head numbers and the total number of network nodes, r is the current round number; G is the cluster node collection with the exception of the cluster head in the last rounds of 1/p. Then on the basis of the intensity of obtained information, the cluster lead node broadcasts the message from its cluster head to the whole network, each node decides to include which cluster and react to the corresponding cluster head.

Suggests a new CODA clustering algorithm mitigate the energy depletion unbalance induced by a distance from the drain. CODA splits the entire network into a few groups depending on the distance from the base station and the routing plan, and each group consists of its cluster members and nodes. In the single-hop case with clustering, other clusters are created farther away from the base station. The network lifespan and dissipated energy have the same risk for the whole network—the clustering scheme concerning the routing technique [12].

Data acquisition is a core aspect in many wireless sensor networks (WSN) applications. It is unrealistic to guarantee anything in real-world applications since they have different energy consumption; the sensors have the same energy. In addition, sensor redeployment contributes to heterogeneous energy ability. The heterogeneous WSN (EDGA) energy scheme has been proposed for successful data collection in this paper. By minimizing energy usage for network communications and balancing the energy load, EDGA achieved reasonable efficiency in terms of lifespan. EDGA depends on the weighted probabilities of election for each node as a cluster head to help control the heterogeneous power capacities.

Furthermore, EDGA adopts a simple but successful solution to address the issue of field coverage in a cluster, i.e., intra-cluster range. The simulation results show that LEACH, HEED, in terms of network life and data obtained in the heterogeneous energy network, is substantially outdoor the suggested EDGA. EDGA protocol interprets heterogeneous nodes to improve network resilience and stabilization. Since cluster heads are used to reception sensing data, signal processing, and transfer the aggregated data to the next node or base station, cluster heads are more energy than cluster members. Between sensor nodes, the cluster head must be rotated. EDGA operates around LEACH and EACP, given how the cluster heads in the heterogeneous network are chosen optimally.

In CBRP (the latest paradigm called headset), the Clustered-based Routing Hierarchical Routing Protocol (CBRP) consists of an active cluster head and several other associated cluster Heads within a cluster. The Headset Member's role is for monitoring and maintaining the network.

These protocols must be further strengthened or new protocols created. These protocols can be generalized to include over three node types and to have more than two hierarchy tiers [11]. Significant problems/factors in these models, where the heterogeneity of sensor nodes is not just present energy but also in their manufacturing and also in their energy consumption (compression, fusion), etc. (Fig. 2.2).

An interesting research question on protocols of this type is how clusters are developed to maximize energy usage and contemporary communication processes, such as latency. The Contents open problems for future studies are influencing cluster formation and cluster-head coordination. In addition, different protocols for energy efficiency, the focus were imposed. Although all of these protocols sound optimistic, many issues and questions have yet to be overcome. In addition, the aggregation and fusion mechanism of data between clusters is also a significant challenge. Additional theoretical future research on protocol routing involves incorporating sensor networks into wired networks. Although their success in terms of energy consumption is encouraging, more research is needed to resolve issues such as video and imaging sensors' quality of services and real-time applications.

2.4 Conclusion

In the last 15 years, the development of ad hoc routing protocols has been one of the most systematic explorations of a design field in the world of informatics history. Although new protocols have improved earlier ones, a single best" protocol cannot be identified. Fearful deployed situations and metric efficiency combinations in which nearly all of the protocols addressed in this article perform their competitors. We have conducted an exhaustive survey of the heterogeneous network in wireless sensor models. The paper is all about top priority being granted to productive energy usage. Different approaches have been explored using a cluster-based, chain-based method to enhance network service life, operation cost, reliability, and throughput variables.

References

1. I.F. Akyildiz, W. Su, Y. Sankara Subramaniam, E. Cayirci, Wireless sensor networks: a survey. J. Comput. Netw. **38**(4), 393–422 (2002)
2. L. Yong-Min, W. Shu-Ci, N. Xiao-Hong, The architecture and characteristics of wireless sensor networks, in *IEEE International Conference on Computer Technology and Development* (Kota Kinabalu, Malaysia, 2009), pp. 561–565
3. J. Yick, B. Mukherjee, D. Ghosal, Wireless sensor networks survey. Comput. Netw. **52**(12), 2292–2330 (2008)
4. V. Parchuri, A. Durazh, L. Baroli, Energy conscious router protocol for heterogeneous wireless sensor networks, in *SPC, LNCS, Computers and Information Science Communications*, vol. 149 (Springer, 2013), p. 158
5. M. Shaikh Enayet Ullah, The hierarchical routing of cluster protocol for WSN IJCNS, vol. 2, no. 5, in *The IEEE 2005, 16th international databank and expert system workshop*, ed. by G. Rashed, M. Hasnat Kabir (2010)
6. H. El Alami, A. Najid, SEFP: a new routing approach using fuzzy logic for clustered heterogeneous wireless sensor networks. Int. J. Smart Sens. Intell. Syst. **8**(9), 2286–2306 (2015)

7. T.S. Barman, S. Maitra, D. Giri, Secure cluster-based energy-efficient routing in wireless networks, in *Information Technology and Applied Mathematics, Smart Systems Advancements and Computers*, vol. 699 (Springer, Singapore, 2019), pp. 23–40
8. J.-S. Lee, T. Kao, An improved three-layer low-energy adaptive clustering hierarchy for wireless sensor networks. IEEE Internet Things **3**, 951–958 (2016)
9. M. Saidu, E.N. Onwuka, M. Okwori, A. Umar, An enhanced LEACH routing algorithm for energy conservation in a wireless sensor network. Int. J. Wirel. Microw. Technol. **6**, 59–71 (2016)
10. A. Chithra, R. Shantha, A Novel 3—Level Energy Heterogeneity Clustering Protocol with Hybrid Routing for a Concentric Circular Wireless Sensor Network, vol. 20, Issue 2 (Springer—Cluster Computing, 2017)
11. K. Haseeb, N. Abbas, M.Q. Saleem, O.E. Sheta, K. Awan, N. Islam et al., RCER: reliable cluster-based energy-aware routing protocol for heterogeneous wireless sensor networks. PLoS One **14**(9), e0222009 (2019)
12. B.M. Khan, R. Bilal, R. Young, Fuzzy-TOPSIS based cluster head selection in mobile wireless sensor networks. J. Electr. Syst. Inf. Technol. **5**(3), 928–943 (2018)
13. M. Sharma, K. Sharma, An energy-efficient extended LEACH (EEE-LEACH), in *2012 International Conference on Communication Systems and Network Technologies* (2012), pp. 377–382. https://doi.org/10.1109/CSNT2012.88
14. S. Mummana, K.N. Rao, An optimized multi sink network for optimal data aggregation in wireless sensor network using invasive weed optimization. J. Adv. Res. Dyn. Control Syst. **10**(04-Special Issue) (2018)

Chapter 3
Sentiment Analysis "Using SVM, KNN and SVM with PCA"

Praveen Verma, Tushar Bhardwaj, Abhay Bhatia, and Mohd Mursleen

Abstract There is a lot of scope of sentimental analysis in the field of natural language processing. Sentiment analysis gives the ability to a programmable system to analyse the sentiment behind some natural language behaving like a human. It does not make the system a human but the ability to understand sentiments behind some natural language that humans use to communicate with each other can be analysed via a computer or system. Sentiment analysis is an excellent way to discover people's views about places, architecture, film, brand, book, story, news, novel and restaurant reviews, etc. In this paper, we compared various sentimental analysis techniques to classify the restaurant reviews as positive or negative sentiments to find the most suitable one to use to achieve maximum accuracy in terms of classification between positive or negative reviews and also automated the approach. This greatly helps the owners of the restaurants to have a clear view of what the customers like about the restaurant and its service and the negative part too. Necessary steps can then be taken by the owner and staff to improve through those reviews. Analyzing every individual review by customers is a hectic task and thus automating this process will reduce costing of labor as well as time. We have used different techniques and methodologies for analyzing the sentiments of restaurant reviews using Support Vector Machine (SVM), K-Nearest Neighbors (KNN), and Support Vector Machine with PCA (Principal Component Analysis). And we achieved an accuracy of about 96% through the SVM model beside other classification models.

P. Verma (✉) · A. Bhatia · M. Mursleen
Department of Computer Science, Roorkee Institute of Technology, Roorkee, Uttarakhand, India
e-mail: praveenverma.cse@ritroorkee.com

A. Bhatia
e-mail: drabhay.cse@ritroorkee.com

M. Mursleen
e-mail: mursleen.cse@ritroorkee.com

T. Bhardwaj
Applied Research Center, Florida International University, Miami, USA

T. Bhardwaj et al. (eds.), *Artificial Intelligence in Cyber Security: Theories and Applications*, Intelligent Systems Reference Library 240,
https://doi.org/10.1007/978-3-031-28581-3_5

Keywords Machine learning · Natural language processing · Scikit-learn · SVM · KNN · PCA

3.1 Introduction

Nowadays, People's view plays a vital role in deciding the future of a business, group, organization as well as the Government and nation growth. People's Opinion, like what they like, whatnot about something is very important to take further actions on it. Even elections are a part of people's Opinions and with the modern technology and rich amount of information available through web media, collecting public Opinion (from Twitter, reviews, Reddit, other social media, etc.) is not a big task anymore and hence the collected data can then be easily analyzed for extraction of the public opinions with their emotions included in it (Fig. 3.1).

In the above described work, we have started with collection of restaurants reviews from internet and have tried to imply our machine learning based classifiers in such a way so that the best model for prediction of positive as well as negative reviews can be find easily. We have collected around on thousand labelled restaurant reviews from the internet and used our approaches of classifying the restaurant reviews using Support Vector Machine, K-Nearest Neighbors and Support Vector Machine with Principal Component Analysis.

There are also a lot of other applications too of Sentiment analysis for discovering off the public opinion about books, films, brands, stories, same groups, places, technology, government policies and many others.

3.2 Literature Review

In [1], author suggested the three levels of sentiment analysis on which basis it has been investigated. These levels are document level, sentence level and aspect level. Document level helps in classifying whether the whole document showing the

Fig. 3.1 General image of SA

positive, negative or neutral sentiment. It was assumed that the whole documents is showing the sentiment for the individual entity. In sentence level of analysis, the major task is to analyses each sentence whether it expressed a positive, negative, or neutral opinion. This level of analysis is closely related to subjectivity classification, which distinguishes objective sentences that express factual information from subjective sentences that express subjective views and opinion. These both level of analysis, Document level and the sentence level are unable to classify the exact opinion of the people what exactly they liked and did not like. The third level of analysis, Aspect level do more exact analysis than other two analysis levels. This level does not consider the language constructs like paragraphs, sentences, clauses or phrases, aspect level directly search for the opinion itself.

In [2], classification of sentiment analysis, Machine learning strategies are used to handle unstructured data or big data. These machine learning algorithms work by training with a training data set, validating on the training data set and then applying these on to the actual data set. We use these algorithm for some particular input data with known output and then applying these algorithms for unknown data.

In [3], authors uses the Support Vector Machine for classification of the scenic data for sentiment analysis. Support Vector Machine is a non-probabilistic classifier that require a large amount of training data set. The classification is done in a 1-D hyper plane to classify the points. It separates the points that belongs to the different classes and arrange them in classes. This process of drawing the boundaries between the points defines the decision boundary that in turn draw a decision plane. Decision plane helps in placing an object to a class with which it relates itself more closely.

K-Nearest Neighbour is a supervised training based machine learning algorithm. The classification is done on the basis of the similarity of an object with the nearby objects available in the vector space. KNN algorithm produces the highly associated classes. The classification is depend on the distance measure. Further, in weighted k-Nearest Neighbour method, a weight is assigned to the elements of training data set and then used this weighted element to measure the sentiment of text [4]. This measurement is done by using the following mathematical equation.

$$\text{Positive Score} = \frac{1 \sum j \ \text{Score(pos)} + 1 \sum k \ \text{Score(neg)}}{1 \sum s \text{Maximum Score}}$$

where s = j + k, i.e. Count of both positive and negative together. In weighted KNN method, tokenization of the sentences is done and then stop phrase is removed from the tweets they have fetched. Two parse algorithm have been proposed by the authors of [4]. In first parse, a positive score is assigned to each reviews. This will be inputted for further parsing and a neutral score is assigned. By using this algorithm, the output is generated in the form of a file that consist of review ID and its positive score.

Fig. 3.2 Sentimental analysis

3.3 Sentiment Analysis (SA)

Sentiment Analysis is about analysing and understanding Sentiments or Opinions of Some natural language [5] mostly in the form of some writing (or written text). It analysis the written text and then determines whether it is positive, negative, or neutral [6]. By making the machine learned about reviews and their labels, it will be able to classify the new reviews [7] (Fig. 3.2).

3.4 Uses of Sentiment Analysis

(1) *Social Media*

Social media is as the major platform that people use to share their opinions in regard of sentiments via the medium like posts, tweets, comments and so on. On an average of around 6000 tweets are shared on social media every second [8]. In which people share their opinion about different contents like movies, business, government policies, etc. with the help of Sentiment Analysis, we can determine whether those reviews, saying, comments and the public opinion are falling in which type of category and it can be done by using the facial expressions too in the form of clip arts too.

The social media is transforming the way we communicate, the way we form relationships, the way we connect to each other, the way we live and work. Here are some figures that give an idea about the frantic pace in which the social media phenomenon is growing: 1.43 billion people worldwide visited a social networking site in 2012; nearly 1 in 8 people worldwide have their own Facebook page 2; 3 million new blogs come online every month; and 65% of social media users said they use it to learn more about brands, products and services.

In Early of year 2007 the researchers and analysts started a new task of monitoring and noticing importance and value of social media and semantic analysis.

(2) *Customer Feedback*

Feedback is the best form of information that a customer provides you based on their personal experiences about the likes and dislike in regard of some products or services. We can get to this with an example that is suppose a customer 'x' visits a restaurant and orders some food but he didn't like the food so the reviews given by that customer

> The food was not that good as expected.

With the help of sentiment analysis, we can determine whether the Feedback is positive or negative. Also, the Sentiment Analysis can be used to process a large amount of customer feedbacks which results into reduction of the labor cost and saving a lot of time.

(3) *Government Organization*

Public sentiment, opinion is very important if we talk about a Government organization. The majority of Government Organizations throughout the world use Sentiment Analysis as a tool to understand public opinion about their policies which is very important for proper administration. The election results can also be predicted from political posts [9]. As established in the facts presented in last section, social media presents itself as a '*big data*' source of citizen voice. If Government agencies can constantly keep a tab on pulse of its citizens, it can pave the way for better governance. Social sentiment analysis can be a very useful tool to achieve the same. Government would be interested to answer question regarding the interest of the people of their country or federations as well as getting proper reviews of their residency.

(4) *Market Research*

One of the Trending Application of Sentiment Analysis is market research to follow up and compete with other businesses. It helps to find the trend with your market effort and the market effort of competitors. Sentiment Analysis helps to analyse the value and work of competitor products in the market and decide what we are lacking in comparison with competitor's businesses. With this business can understand the strength and weakness and can further take necessary future actions to increase their strength and come up with a solution to their weakness to improve their marketing services and decision policies [10].

Analyse the emotion around these discussions to make sure it's something you want to be included in. Also, if you want to enter a new market or release a new product—sentiment analysis should be a part of your market research to inform your decisions. It also tells about the marketing campaign and movie around the marketing research topics and what hot works can be done in the market for latest working and implementation these to the markets (Fig. 3.3) [11].

(5) *Brand analysis*

Brand analysis is a process of marketing that helps us to evaluate your company's branding strategies and methods. Generally branding involves the company's profile, personality and identity. Generates the message that is sent to the customers and potential clients. Brand analysis is a set of steps and strategies that you can use to assess your company's current branding strategies. We can use this analysis to find your marketing strengths and identify areas of improvement.

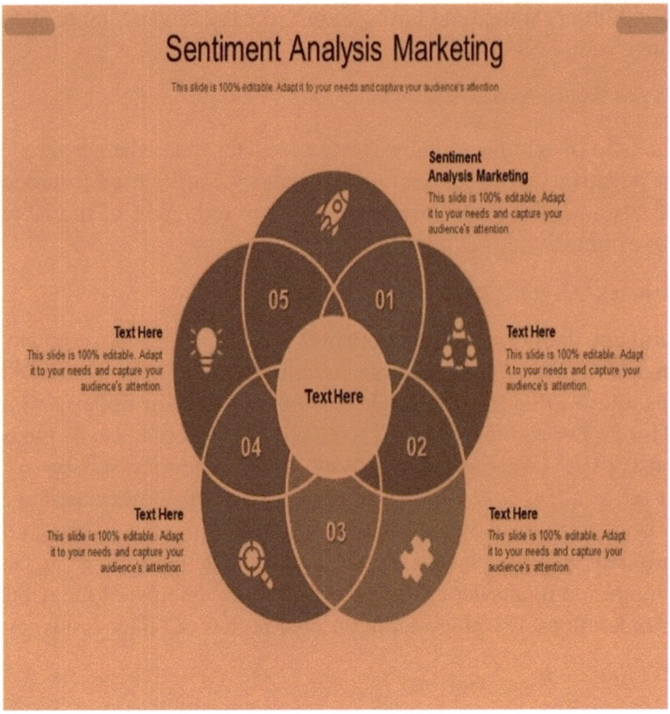

Fig. 3.3 Market research process

Brand analysis and continuous monitoring are one of the most crucial for a business either online or offline. Brand reputation and its image are very important to keep track of its strength and weaknesses. Sentiment analysis plays a very important role in this. It can help a business by processing a large number of tweets, blog reviews to understand public opinion about a brand. And as we know doing this work manually is not a good approach. Therefore, sentiment analysis comes to our help by automating the approach as it can process and analysis a large number of feedbacks and reviews and reduce labor cost and labor. It includes the following on which basis it works:

- Goals for branding.
- Objective.
- Results for the generated objectives.
- Feedbacks (internal basis).
- Analysis of Market.
- Summarized company profile.
- Analysis for competitive part.

3.5 Data Preprocessing

Pre-processing the data is the process of cleaning and preparing the text for classification [12]. Text data contains lots of features and noise and therefore it becomes very important to first clean the data before passing it to the model. In addition, many word in the text are irrelevant and unnecessary. Keeping those words makes the classification process harder.

In our approach, we performed various techniques in order to clean our dataset. We first performed lower characterization to make all the reviews small so we now have less number of dimensions. We then also removed some non-useful words like a, an, the, etc. from the datasets using Natural Language Toolkit (NLTK) library. This process of removing non useful words from the context is also called stop-word elimination. We then also performed stemming, an approach to reduce a word to its stem or root form [13] and tokenization.

3.6 Sentiment Analysis—A Machine Learning Approach

Machine learning is the study of computer algorithms that allow computer programs to automatically improve through experience [14, 15] (Fig. 3.4).

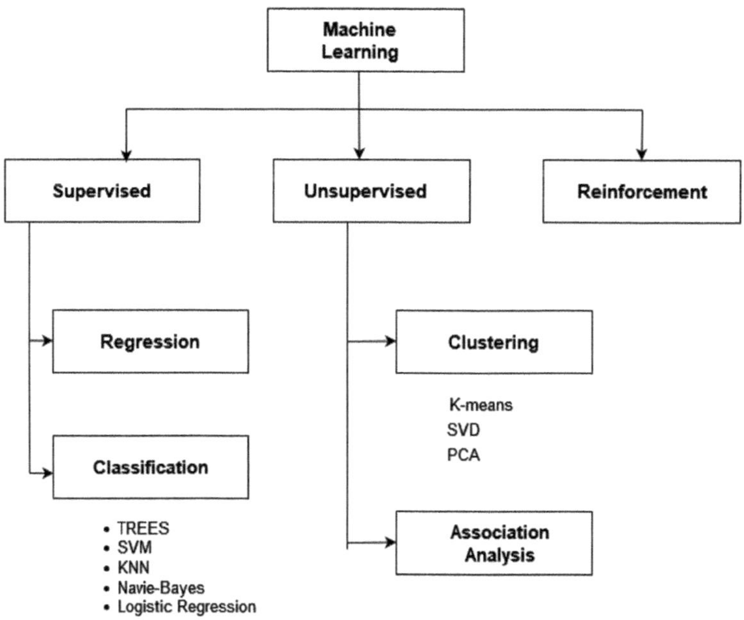

Fig. 3.4 General structure of machine learning [15]

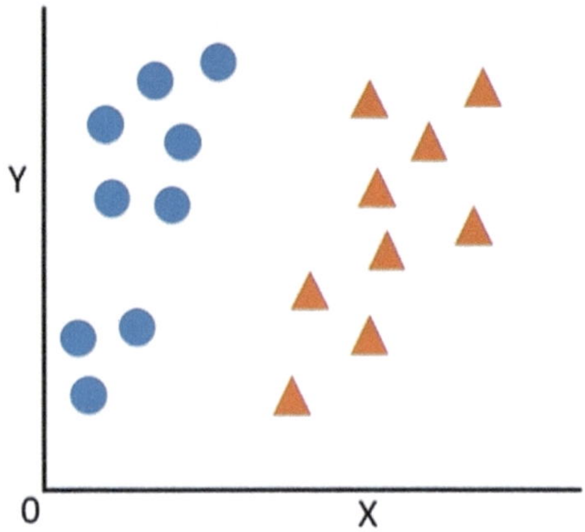

Fig. 3.5 Labeled diagram

In this paper, our main focus is to show the comparison between Support Vector Machine (SVM), K-Nearest Neighbors (KNN) classification technique, and SVM with Principal Component Analysis (PCA) using restaurant reviews.

A. **Support Vector Machine**

Support Vector Machine is a supervised classification technique that is used to distinctly classify the data points to find a hyper plane in N-dimensional space (N denotes the number of features). Apart from the classification task it can be applied to regression problem and outlines detection.

A hyper plane can be explained as a decision boundary that differentiates the two classes in SVM. A data point falling on either side of the hyper plane can be attributed to different classes. The dimension of the hyper plane depends on the number of input features in the dataset. If we have 2 input features the hyper-plane will be a line. Likewise, if the number of features is 3, it will become a two-dimensional plane (Figs. 3.5, 3.6 and 3.7) [16–18].

To classify the data points into different categories SVM constructs a hyper plane or set of hyper plane, which can then be used for classification and regression tasks [19]. A hyper plane is simply decision boundaries that help to classify the data points whether the data points are falling under class A (blue circle) or class B (orange triangle). There can be multiple hyper planes possible but the hyper plane (in linear data) whose distance to the nearest element of each data point is the largest is considered as the best hyperplane [6] and the distance is called margin. If we select a hyperplane with a low margin then there are high chances of misclassification. Therefore, we need the best of all the hyperplanes to have a good classification model.

A.1 **Working of SVM**

The basics working of Support Vector Machines can be understood with a simple example. Let's imagine we have two tags: *orange* and *blue*, and our data has two features: *x* and *y*. We want a classifier that, given a pair of *(x, y)* coordinates, outputs

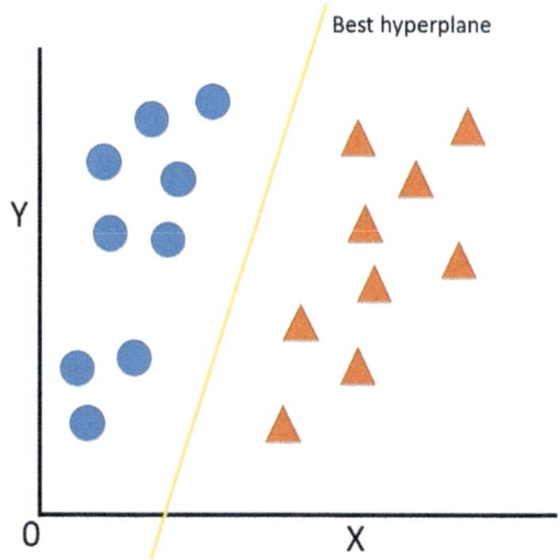

Fig. 3.6 With hyper plane (2D)

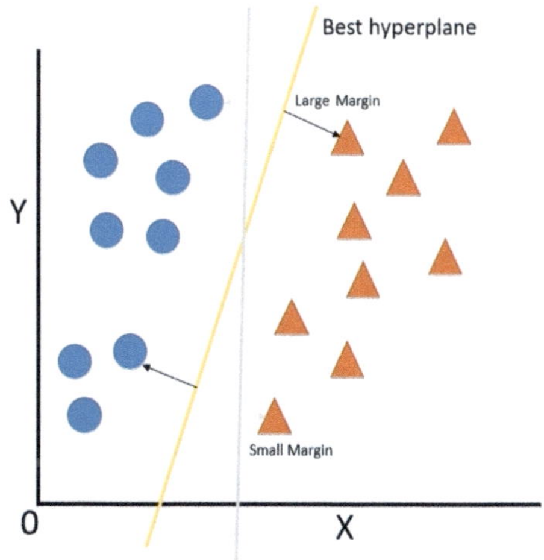

Fig. 3.7 Multiple hyper plane possible

if it's either *orange* or *blue*. We plot our already labeled training data on a plane (Fig. 3.5).

A SVM takes these data points and generates the outputs the hyperplane (2D it's simply a line) that best separates the them up. This line is the **decision boundary**: anything that falls to one side of it we will classify as *blue*, and anything that falls to the other as *orange*.

But, exactly *the best* hyperplane for SVM is the one that maximizes the margins from both the tags. In other words: the hyperplane which resides nearest element whose distance to the tag is the largest one.

A.2 Merits and Demerits associated with it

Merits

- Where the margins are clear the working of it can be seen very well.
- In spaces having high dimensions the effectiveness of it can be seen.
- If the number of samples are less in number than to the dimensions results in increment of its performance.
- Support vectors are used as training point in the decision function which states its memory efficiency.

Demerits

- As its training time is higher so for large amount of data the performance of it is not well.
- If data set is found with more noise than it's really hard to have good performance measure.
- The major one is that it's not efficient in providing probability estimations as for such huge calculations expensive validators are required.

B. K-Nearest Neighbors (KNN)

- Based on Supervised Learning technique K-Nearest Neighbour can be seen as one of the simplest Machine Learning algorithms
- K-NN is an algorithm that is capable of assuming the similarity between the existing cases and newly defined cases.
- After storing all the available data it classifies the new data points mostly based on similarities. Hence it classifies the well suite category of the data.
- It's a non-parametric algorithm as for underlying data no such assumptions are made.
- As SVM it's not a good learner as it does not learn anything from the training set easily which give it a term lazy learner algorithm.
- KNN algorithm at the training phase just stores the dataset and when it gets new data, and then it classifies that data into a category that is much similar to the new data (Fig. 3.8).

B.1 Working of KNN

- **Step-1**: Select the number K of the neighbours.
- **Step-2**: Calculate the Euclidean distance of **K** number of neighbours.
- **Step-3**: Take the K nearest neighbours as per the calculated Euclidean distance.
- **Step-4**: Among these k neighbours, count the number of the data points in each category.
- **Step-5**: Assign the new data points to that category for which the number of the neighbour is maximum.
- **Step-6**: Our model is ready.

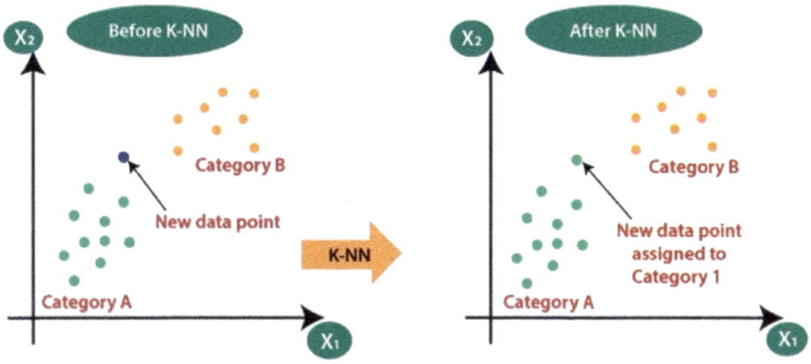

Fig. 3.8 How KNN works

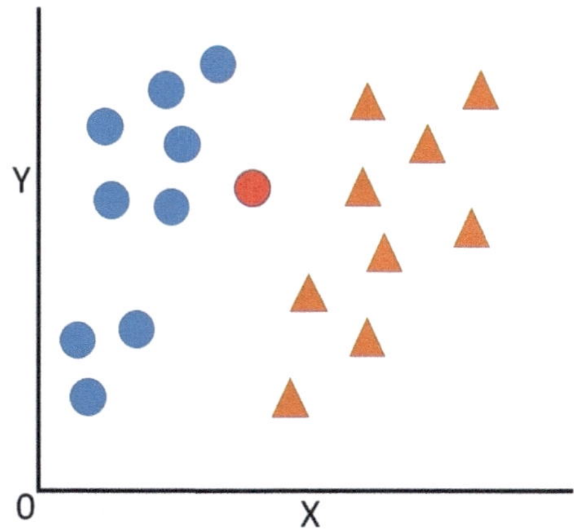

Fig. 3.9 Labelled data

Suppose we select k = 7, then it mean we have to consider 7 nearest neighbors and calculate the distance between all those 7 points with the new data points (Fig. 3.9).

From Fig. 3.10, we can conclude that out of 7 nearest neighbors from the new data points. 5 belong to class A (blue circle) and 3 belongs to class B (orange triangle).

The KNN algorithm, says the new data points belong to the class having the maximum number of nearest neighbors. Therefore, the new data points belong to the class A (blue circle).

(1) *Best Value of K:* To select the best value of K we consider it to be odd. If we consider an even value for k, let say k = 6, then there is a chance of getting an equal no. of neighbors for each class. Therefore, an odd k value is said to be the best value of K.

Since guessing the best value of K does not always prove to be the best choice. So, how do we select the best value of K? To select the best value of K we can have a subset of the actual data and create a K-NN classifier with a different value of K's using a loop, and draw a line chart for accuracy metric.

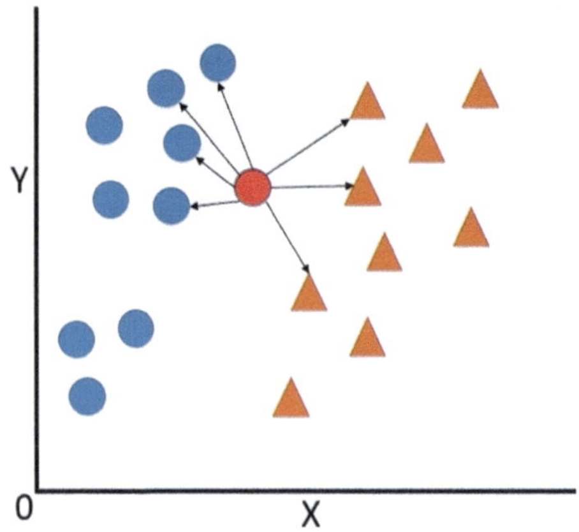

Fig. 3.10 Working of KNN with a new data point

From the above Fig. 3.11, we can say that K = 7 seems to be an ideal/best value for K.

B.2 **Significance of K**

Specifically, the KNN algorithm works in the way:

Finds a distance between a query and all examples of data, select the particular number of examples (say K) nearest to the query, and then decide

- the most frequent label if using for the classification based problems,

  OR

- the averages the label if using for regression-based problems

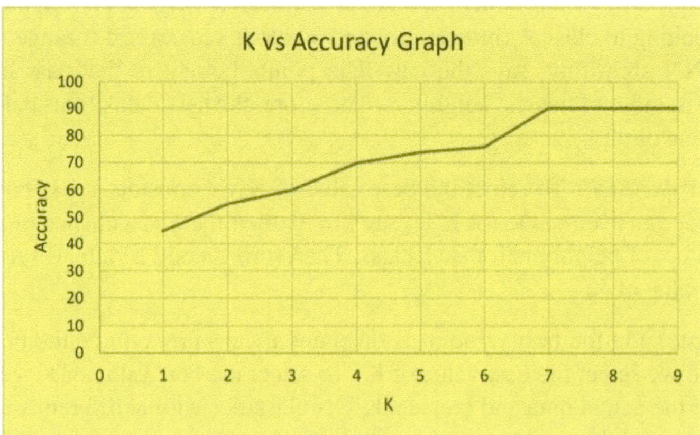

Fig. 3.11 K versus accuracy graph

Therefore, the algorithm completely depends upon the number of K, such that.

- Value of k—bigger the value of k increases confidence in the prediction.
- Decisions may be skewed if k has a very large value.

B.3 Pseudo code

a. Desired data to be loaded.
b. Value of k must be chosen.
c. Starting from first step the repetition will be done till all the described points for training are covered for predicting the classes.
d. By using Euclidean distance, the calculation over the training data points and calculate the distance between the data point whose class is distance between the predicted points is done.
e. We have to arrange a increment order of the distance for proper arrangement of the distances.
f. By assuming positive value for k value then start filtration process for sorting of the lowest value of k.
g. Now we have achieved the top most values for k distances.
h. Let ka_a represent the points that belong to the ath class among k points.
i. If ka > kb then put x in the class.

B.4 Merits and Demerits associated with it

Merits

- Implementation is very simple and easy.
- The noisy training data robusting can be achieved.
- The effectiveness can be achieved if the available training data is in large amount.

Demerits

- The requirement of high storage.
- Low and slow prediction rate.
- Generally all training data set is stored.
- The algorithm get slower when the number of examples, predictors or independent variables increases.

C. *Principal Component Analysis*

Principal component analysis (PCA) is a multivariate technique that analyses a data table in which observations are described by several inter-correlated quantitative dependent variables [20]. Principal Component Analysis is an unsupervised learning dimensionality-reduction technique used to reduce the dimensionality of large data sets. The dimensionality is reduced but the data set still contains most of the information. In a nutshell, the idea of using PCA is to reduce the no. of variables in a large data set while still preserving most of the information in exchange for losing little accuracy.

Working with large data sets is a resource-hungry task and requires high/heavy computation power. But since working with small data sets is not only just easy but also requires a smaller number of resources and computation power.

PCA drops the least important variables in the data sets in such a way that important variables are still available and each variable after PCA is all independent of each other. Using PCA can be a challenging task as it removes variables from our original dataset and if it is not done properly, then we might end up losing important variables also may affect our accuracy. Therefore, having proper knowledge of PCA is important.

C.1 Steps required in PCA

- **Standardization**

  The aim of this step is to standardize the range of the continuous initial variables so that each one of them contributes equally to the analysis. More specifically, the reason why it is critical to perform standardization prior to PCA, is that the latter is quite sensitive regarding the variances of the initial variables. Mathematically it can be done as

$$z = \frac{\text{Value} - \text{Mean}}{\text{standard deviation}}$$

- **Covariance matrix computation**

  To see the relationship between the variables or how the data set vary from mean w.r.t each other.

  Consider a symmetric matrix p*p (dimensions = p) and data set of 3 dimensional covariance matrix having variables x, y and z.

$$cov(x, x)\, cov(x, y)\, cov(x, z)\, cov(y, x)\, cov(y, y)\, cov(y, z)\, cov(z, x)\, cov(z, y)\, cov(z, z)$$

- **Compute the eigenvectors and eigenvalues of the covariance matrix to identify the principal components**: Eigenvectors and eigenvalues are the linear algebra concepts that we need to compute from the covariance matrix in order to determine the principal components of the data.

  Principal components are new variables that are constructed as linear combinations or mixtures of the initial variables. These combinations are done in such a way that the new variables (i.e., principal components) are uncorrelated and most of the information within the initial variables is squeezed or compressed into the first components.

  An important thing to realize here is that, the principal components are less interpretable and don't have any real meaning since they are constructed as linear combinations of the initial variables.

C.2 **PCA principle components**

- The first principal component accounts the largest possible variance for any desired data set.
- The second one works in same way to account the next highest variance in uncorrelated conditions.

C.3 **Feature Vector**

In this step we choose which component is to be kept and which one to be discarded.

It's simply a matrix having eigenvector of the components that are decided.

C.4 **Recast the Data Along the Principal Components Axes**

We do not make any changes on the data, just select the principal components and form the feature vector, but the input data set remains always in terms of the original axes.

The aim of this step is using feature vector which is formed with the help eigenvector of covariance matrix in such a way that it reorient data from original axis to the one representative component.

3.7 Results and Performance Analysis

The three classifier have been individually tested on the test dataset of restaurant reviews and the accuracy is calculated and then compared to find the best model to work with sentiment analysis tasks. The results are evaluated using accuracy score and confusion matrix.

A. Performance of Support Vector Machine

Through SVM we got an accuracy of about $96.4285 \approx 96.43\%$. From Table 3.1, we can see that among 70 bad reviews, it was able to predict 67 review correctly and among 70 good reviews, SVM predicted 68 review correctly (Fig. 3.12).

B. Performance of KNN

Through KNN we got an accuracy of about $78.5714 \approx 78.57\%$. From Table 3.2, we can see that among 70 bad reviews, it was able to predict 59 reviews correctly and among 70 good reviews, SVM predicted 51 reviews correctly (Fig. 3.13).

Table 3.1 Performance analysis SVM model

	Bad (0)	Good (1)
Actual	70	70
Predicted sentiment	69	71
Right predicted	67	68
Wrong predicted	3	2

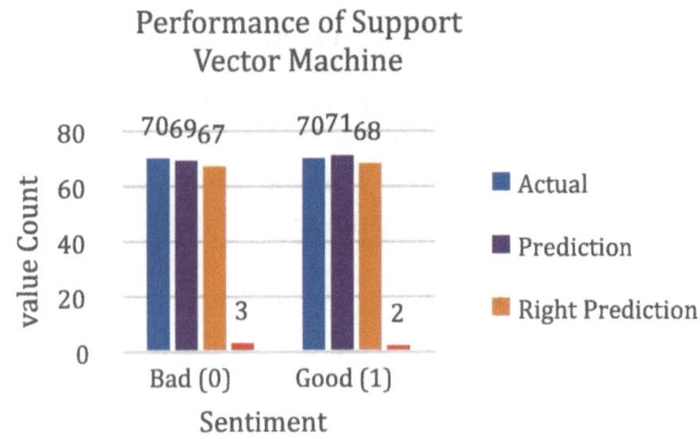

Fig. 3.12 Graphical representation of SVM performance [17, 18]

Table 3.2 Performance analysis KNN model

	Bad (0)	Good (1)
Actual	70	70
Predicted sentiment	78	62
Right predicted	59	51
Wrong predicted	11	19

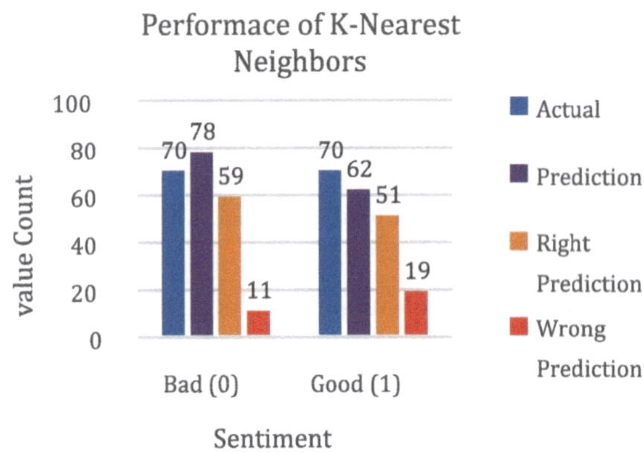

Fig. 3.13 Graphical representation of KNN performance [17, 18, 21]

C. Performance of SVM with PCA

Through SVM with PCA, we got an accuracy of about 88%. From Table 3.3, we can see that among 49 bad reviews, it was able to predict 45 review correctly and among 51 good reviews, SVM predicted 43 review correctly (Fig. 3.14).

Table 3.3 Performance analysis SVM with PCA model

	Bad (0)	Good (1)
Actual	49	51
Predicted sentiment	53	47
Right predicted	45	43
Wrong predicted	4	8

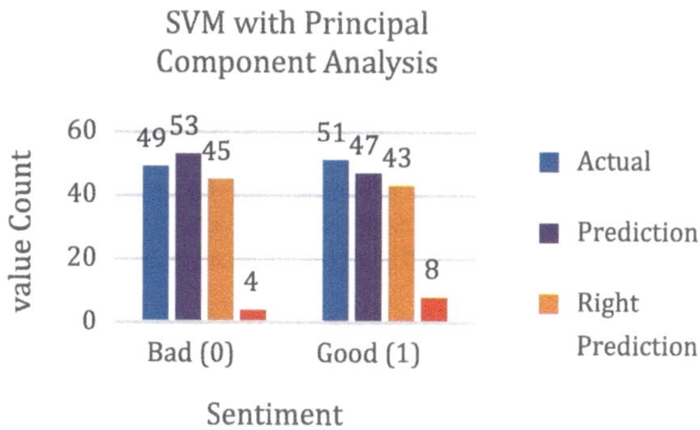

Fig. 3.14 Graphical representation of SVM with PCA performance [22, 23, 24, 25]

3.8 Conclusion

In this paper, we came to find that the Support Vector Machine classifier performed best besides other models for sentimental analysis.

Support Vector Machine can be used for sentimental analysis for tweets, customer feedback, brand analysis, etc. On our test data using the SVM model, we got an accuracy of about 96%.

In this work, we implemented an automated model of sentiment analysis for the classification of restaurant reviews [26, 22]. As reviewing a huge amount of reviews can be a tough task, thus automating the approach and using sentiment analysis proved to be a good choice. By using this model, the owner of such businesses will easily be able to get an idea about their services being offered to the customers. Also, the customers will be able to know about their opinions of other people about the restaurant and make their judgment to go there or not. And such a system will be very valuable to the other domains also that require opinion mining.

References

1. N.S. Joshi, S.A. Itkat, A survey on feature level sentiment analysis. (IJCSIT) Int. J. Comput. Sci. Inf. Technol. **5**(4), 5422–5425 (2014)
2. Y. He, Incorporating sentiment prior knowledge for weakly supervised sentiment analysis. ACM Trans. Asian Lang. Inf. Process. **11**(2)

3. N. Veeranjaneyulu, A. Raghunath, B.J. Devi, V.N. Mandhala, Scene classification using support vector machines with Lda. J. Theor. Appl. Inf. Technol. **63**(3) (2014)
4. A. Srivastava, M.P. Singh, Supervised SA of product reviews using weighted k-NN algorithm, in *2014 11th International Conference on Information Technology*
5. G. Angiani, et al., A comparison between preprocessing techniques for sentiment analysis in Twitter, in *KDWeb* (2016)
6. Sentiment Analysis: A Definitive Guide [Online]. https://monkeylearn.com/sentiment-analysis/
7. M. Hu, B. Liu, Mining and summarizing customer reviews, in *Proceedings of the Tenth ACM SIGKDD International Conference on Knowledge Discovery and Data Mining* (ACM, 2004)
8. Twitter Usage Statistics [Online]. https://www.internetlivestats.com/twitter-statistics/
9. C. Spoorthi, P.R. Kumar, M.J. Adarsh, Sentiment analysis of customer feedback on restaurants. Int. J. Eng. Res. Technol. (IJERT) ICRTT—2018 **6**(15)
10. E.M.G. Younis, Sentiment analysis and text mining for social media microblogs using open source tools: an empirical study. Int. J. Comput. Appl. **112**(5), 44–48 (2015)
11. R.A. Laksono, K.R. Sungkono, R. Sarno, C.S. Wahyuni, Sentiment analysis of restaurant customer reviews on tripadvisor using naïve bayes (2019), pp. 49–54. https://doi.org/10.1109/ICTS.2019.8850982.
12. E. Haddi, X. Liu, Y. Shi, The role of text pre-processing in sentiment analysis. Procedia Comput. Sci. **17**, 26–32 (2013). https://doi.org/10.1016/j.procs.2013.05.005
13. D. Sharma, Stemming algorithms: a comparative study and their analysis. Int. J. Appl. Inf. Syst. **4**(3), 7–12 (2012)
14. Machine Learning Definition | Tom M. Mitchell| McGraw-Hill Science/Engineering/Math; (March 1, 1997), Page 1 | http://www.cs.cmu.edu/afs/cs.cmu.edu/user/mitchell/ftp/mlbook.html
15. E. Alpaydin, *Introduction to Machine Learning* [Online]. MIT Press
16. N. Zainuddin, A. Selamat, Sentiment analysis using support vector machine, in *I4CT 2014—1st International Conference on Computer, Communications, and Control Technology, Proceedings* (2014), pp. 333–337. https://doi.org/10.1109/I4CT.2014.6914200
17. M. Rezwanul, A. Ali, A. Rahman, Sentiment analysis on Twitter data using KNN and SVM. Int. J. Adv. Comput. Sci. Appl. **8** (2017). https://doi.org/10.14569/IJACSA.2017.080603.
18. S. Kaur, G. Sikka, L.K. Awasthi, Sentiment analysis approach based on N-gram and KNN classifier, in *2018 First International Conference on Secure Cyber Computing and Communication (ICSCCC)* (Jalandhar, India, 2018), pp. 1–4. https://doi.org/10.1109/ICSCCC.2018.8703350
19. Support-vector machine [Online]. https://en.wikipedia.org/wiki/Support-vector_machine
20. B. Keith Norambuena, E. Lettura, C. Villegas, Sentiment analysis and opinion mining applied to scientific paper reviews. Intell. Data Anal. **23**, 191–214 (2019). https://doi.org/10.3233/IDA-173807
21. G. Vinodhini, R.M. Chandrasekaran, Sentiment classification using principal component analysis based neural network model, in *International Conference on Information Communication and Embedded Systems (ICICES 2014)* (Chennai, India, 2014), pp. 1–6. https://doi.org/10.1109/ICICES.2014.7033961
22. G. Vinodhini, R. Chandrasekaran, Opinion mining using principal component analysis based ensemble model for e-commerce application. CSIT **2**, 169–179 (2014). https://doi.org/10.1007/s40012-014-0055-3
23. S. Yi, X. Liu, Machine learning based customer sentiment analysis for recommending shoppers, shops based on customers' review. Complex Intell. Syst. **6**, 621–634 (2020). https://doi.org/10.1007/s40747-020-00155-2
24. The new eye of government: citizen sentiment analysis in social media. https://aclanthology.org/W13-4204.pdf
25. Principal component analysis: a review and recent developments Ian T. Jolliffe and Jorge Cadima. https://doi.org/10.1098/rsta.2015.0202
26. J. Jotheeswaran, M. Sudhanan, R. Loganathan, Feature reduction using principal component analysis for opinion mining. Int. J. Comput. Sci. Telecommun. **3**, 118 (2012)

27. X. Fang, J. Zhan, Sentiment analysis using product review data. J. Big Data **2**, 5 (2015). https://doi.org/10.1186/s40537-015-0015-2
28. Data Pre-processing in text [Online]. https://towardsdatascience.com/nlp-forbeginners-cleaning-preprocessing-text-data-ae8e306bef0f
29. Y. Dang, Y. Zhang, H. Chen, A lexicon-enhanced method for sentiment classification: an experiment on online product reviews. IEEE Intell. Syst. **25**(4), 46–53 (2010). https://doi.org/10.1109/MIS.2009.105
30. M. Adnan, R. Sarno, K.R. Sungkono, Sentiment analysis of restaurant review with classification approach in the decision tree-J48 algorithm, in *2019 International Seminar on Application for Technology of Information and Communication (iSemantic)* (Semarang, Indonesia, 2019), pp. 121–126. https://doi.org/10.1109/ISEMANTIC.2019.8884282
31. M. Govindarajan, Sentiment analysis of restaurant reviews using hybrid classification method (2014)
32. S. Mishra, U. Sarkar, S. Taraphder, S. Datta, D. Swain, R. Saikhom, S. Panda, M. Laishram, Principal component analysis. Int. J. Livestock Res. (2017). https://doi.org/10.5455/ijlr.20170415115235
33. M.Z. Asghar, A. Khan, S. Ahmad, F.M. Kundi, A review of feature extraction in sentiment analysis. J. Basic Appl. Sci. Res. **4**(3), 181–186 (2014)
34. A. Shathik, K.P. Karani, A literature review on application of sentiment analysis using machine learning techniques 2581–7000 (2020). https://doi.org/10.5281/zenodo.3977576
35. R. Feldman, Techniques and applications for sentiment analysis. Commun. ACM **56**(4), 82–89 (2013)

Chapter 4
Security Attack Detection Using Machine Learning and Deep Learning

Tarun Jain, Nityaa Kalra, Deepak Panwar, and Aditya Sinha

Abstract Data has been increasing exponentially over the past few years. This has introduced a variety of anomalies and security threats in the field of networking. Because of this, Intrusion Detection Systems (IDS) have become popular. An IDS is a network trafficking monitoring system that detects any suspicious activity and alerts the administrator when it is found. An IDS is generally prone to false alarms. Therefore, there is a need to fine-tune them. Using Machine Learning and Deep Learning techniques, one can build such systems which are accurate in detection of malicious activities. Previously, KDDCUP99 dataset has been used for most of the research purposes. However, the dataset is old and doesn't justify the intrusions and malicious activities that occur now. Therefore, we have made the use of UNSW-NB15dataset (Moustafa and Slay, 2015 Military communications and information systems conference (MilCIS), 2015) which is an upgraded version of KDDCUP99 dataset and is more uniform and balanced. We have used various machine learning and deep learning algorithms like Logistic Regression (LR), Decision Trees (DT), Random Forest (RF), Naïve Bayes (NB), Support Vector Machine (SVM) and Artificial Neural Networks (ANN) to classify whether it was a network intrusion or not. It has been found that SVM, DT, ANN and NB gave a hundred percent accuracy whereas the Random Forest model gave an accuracy of 99.9%.

4.1 Introduction

With advancing technology and expansion in the use of Internet, network security issues have come into existence. Various hackers try to break into the networking systems resulting in a security breach. The tradition security systems are ineffective when it comes to prevention of network intrusions in the ever-expanding cyberspace. This demands for a system that is far more robust and learns from the data provided. This is where ML based Intrusion Detection Systems play a crucial role in the

T. Jain · N. Kalra · D. Panwar · A. Sinha (✉)
Manipal University Jaipur, Jaipur, Rajasthan, India
e-mail: adityasinha.1988@gmail.com

T. Bhardwaj et al. (eds.), *Artificial Intelligence in Cyber Security: Theories and Applications*, Intelligent Systems Reference Library 240,
https://doi.org/10.1007/978-3-031-28581-3_6

computer network security. An Intrusion Detection System in mainly divided into three broad categories [2–4]:

1. *Host-based IDS*: It works on independent device on which is installed and is concerned with detection of malicious activities, monitoring of operating system files and informs the administrator about the possible threats to the environment.
2. *Network IDS*: A Network IDS works on the all the devices present in the environment instead of a single device and examines the incoming and outgoing traffic in the entire subnetwork. In case any anomaly is detected, the admin in informed.
3. *Hybrid IDS*: A hybrid IDS is a combination of *Host-based* and *Network* IDS [5]. They are more effective and effective as compared to the standard IDS because they combine the hosts and network information.

All these classes can be classified into main approaches namely: *Signature-based* IDS which focuses on detection of any threat based on the data obtained from previously occurred attacks and *Anomaly-based* IDS that detects any unknown activity and flags them as threats and *Hybrid-based* IDS which is a combination of both [6].

4.2 Literature Review

Various scientists and researchers have worked on Intrusion Detection Systems and have tried integrating them with Machine Learning. A brief outline of works of different researchers is given below.

Ahmad et al. [6] used feature selection techniques to reduce data dimensionality. Apart from this, AdaBoost based approach was implemented on Decision Tree classifier. An accuracy of 99.3% was obtained.

Jing et al. [7] used SVM classifier for both binary and multi-class classification of attacks. Three performance metrics namely: accuracy, detection rate and false positive rate were considered while selecting SVM as the most appropriate classifier. An accuracy of 85.99% was achieved for binary classification and 75.77% for multi-class classification.

Sharma et al. [8] implemented various ensemble learning methods like Random Forest, Adaboost, XGBoost and Extra Trees for classification. Major focus has been made on Exploratory Data Analysis along with bagging and boosting methods. The Random Forest classifier gives the highest accuracy of 86.99%.

Bhati et al. [9] used KDDCUP99 dataset for their research. A special focus was made on detection of U2R and R2L attacks through discriminant classification. An accuracy of 98.9% was achieved. It was concluded that their models performed better than the classic ensemble-based approaches.

Choudhary et al. [10] focused on attacks on IoT devices and made use of Deep Neural Networks for identifying these attacks. Three datasets: KDDCUP99, UNSW-NB15 and NSL-KDD were used and accuracies greater than 90% were found in each case.

Naseer et al. [11] trained deep learning models on NSL-KDD datasets and tested on NSLKDDTest+ and NSLKDDTest21 datasets. Various machine learning techniques such as KNN, Decision Trees, Quadratic Discriminant Analysis were also used.

Sarker et al. [12] used tree-based classification on dataset obtained from Kaggle. The researchers have named their approach as '*IntruDtree*' which is much more efficient and robust than the classical machine learning methods and gave an accuracy of 98%.

Ren et al. [13] proposed DO_IDS which is a hybrid data optimization technique consisting of both feature selection and data sampling. *iForest* has been used for elimination of outliers and *Genetic Algorithm* for optimizing the dataset. The Random Forest classifier has been used with GA to give an accuracy of 92.8%.

4.3 Methodology

Figure 4.1 represents the methodology implemented to classify the network security attacks appropriately. A definite set of steps have been used for the same.

4.3.1 Data Set

The data set used is UNSW-NB15, which is an upgraded and well-defined version of the previous KDDCUP99 dataset which was previously used widely for training of Intrusion Detection Systems. This dataset is smaller than the KDDCUP99 dataset but

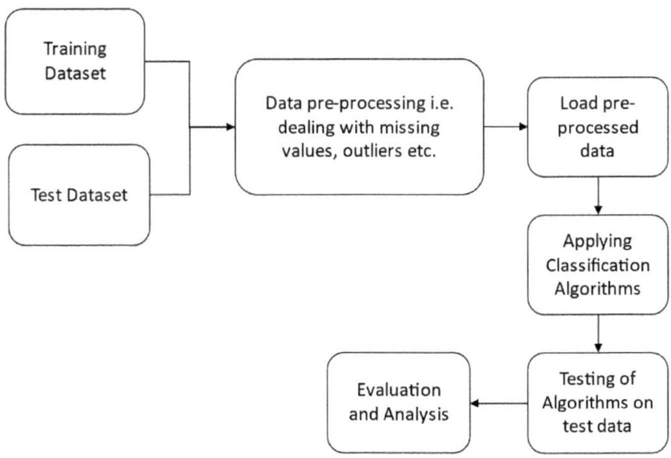

Fig. 4.1 Proposed methodology

Table 4.1 Description of categories of UNSW-NB15 dataset

Sr. no.	Attack type	Description
1	Normal	Natural transaction data
2	Analysis	An attack that invades web applications through emails, ports or web scripts
3	Backdoor	Using backdoor to secure remote access
4	DoS	A malicious attempt to disrupt the computer resources by attacking memory
5	Exploits	An instruction that takes advantage of bugs/errors caused by unintentional behavior on the network
6	Fuzzers	An attack to crash the system by inputting a lot of random data
7	Generic	A technique to clash the block-cipher configuration by using hash functions
8	Reconnaissance	A probe to evade network security controls by collecting relevant information
9	Shellcode	Code is used to exploit software vulnerabilities
10	Worms	A set of virus codes can add to a computer system or other programs

is less redundant and is sufficient to train an IDS efficiently. The dataset consists of 257,673 instances out of which 82,332 instances are used in the training set and the remaining 175,341 are used for testing. A total of 45 features are present. The dataset has 10 targets namely: Normal, Analysis, Backdoor, DoS, Exploits, Fuzzers, Generic, Reconnaissance, Shellcode and Worms. Table 4.1 gives a detailed description of these categories. The binary version of these targets is also available, which is 0 for Normal data transaction and 1 for an attack. Here, we make the use of this binary version to classify whether the attack took place or not.

4.3.2 Data Pre-processing

Even though the dataset is already balanced and doesn't consist of any missing values, the data needs to be scaled and non-contributing features need to be removed. Scaling techniques like *StandardScaler* and *MinMaxScaler* have been used to scale the dataset to obtain the desired results. Since we are working on binary classification, the column '*attack_cat*' has been removed. Apart from this, the '*id*' column has also been removed from the dataset.

4.3.3 *Classification Algorithms*

The dataset has been trained and tested on six algorithms. The detailed description of these algorithms is given below [14, 15].

4.3.3.1 Logistic Regression

When the dependent variable only two possible outcomes, logistic regression is commonly utilised. It is a predictive analysis that demonstrates the relationship between the dependent and independent variables (one or more). It employs the sigmoid function, which maps any real value between 0 and 1. The hypothesis of Logistic Regression is represented by Eq. 4.1.

$$h\theta(X) = \frac{1}{1 + e^{-(\beta_0 + \beta_1 X)}} \; (Hypothesis\; of\; Logistic\; Regression) \tag{4.1}$$

4.3.3.2 Decision Trees

A decision tree is a non-parametric, tree-like model used for classification. Each node in the tree is a test on an attribute and each branch represents the outcome of the previous node. The final leaf node holds the predicted class.

4.3.3.3 Random Forest Classifier

A RFC uses a group of decision trees on various data points. Each decision tree predicts a class. Together, they all work as an ensemble and the class which is predicted the most is the result.

4.3.3.4 Naïve Bayes

A Naïve Bayes classifier is based on the Bayes theorem and assumes that the features are independent of each other. It can build models that give quick predictions and works on the basis of probabilities of data points.

4.3.3.5 Support Vector Machine

A SVM uses a hyperplane which acts as a decision boundary to classify the data points in the data space. Figure 4.2 represents the working of a SVM classifier.

Fig. 4.2 SVM classifier

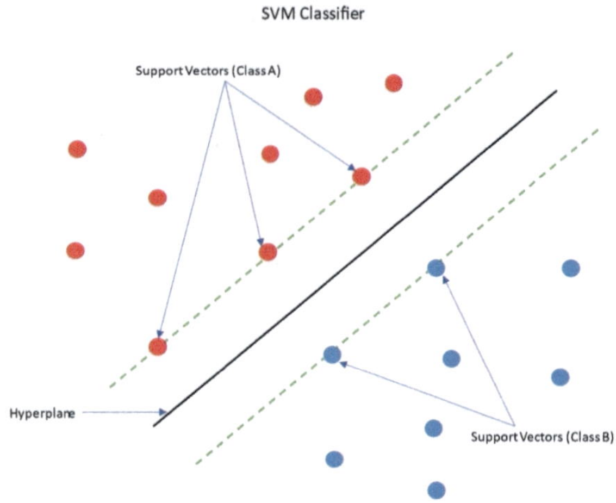

4.3.3.6 Artificial Neural Network

A neural network is a computational model that mimics the operation of nerve cells in the human brain. The first layer is made up of input neurons, which are then followed by hidden layers and a final output layer. The output of the previous layer is used as the input for the next layer. Figure 4.3 depicts the operation of an Artificial Neural Network.

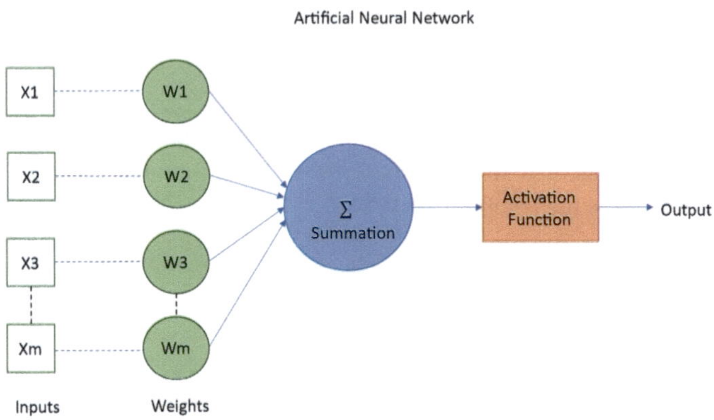

Fig. 4.3 Artificial neural network

4.4 Result and Discussion

While deciding which model works best for the given dataset, it is important to realise that Accuracy, Recall and False Positive Rate are the most important deciding metrics. These metrics can be obtained through confusion matrix.

Evaluating the performance of any machine learning model based on numerous criteria is essential. This aids in determining the best algorithm for the given problem. It is always important to select proper metrics for evaluating the model. A confusion matrix is commonly used to assess various evaluation parameters.

Confusion Matrix

It is an N × N matrix formed using various combinations of the predicted values and the actual values (Table 4.2).

Based on this confusion matrix, we can also calculate the following metrics:

1. *Accuracy*: It is the ratio of the number of correct predictions to the total number of samples.

$$Accuracy = \frac{TP + TN}{TP + TN + FP + FN}$$

2. *Precision*: Precision reveals how many of the accurately anticipated cases were positive.

$$Precision = \frac{TP}{FP + TP}$$

3. *Recall*: It explains how many actual positive values we were able to anticipate correctly. It is equivalent to Sensitivity or Decision Rate.

$$Recall = \frac{TP}{TP + FN}$$

4. *F1-Score*: It is a harmonic mean of recall and precision with a value ranging from 0 to 1. Models with a better F1-score have greater dependability.

$$F1\,score = 2 * \frac{Precision * Recall}{Precision + Recall}$$

Table 4.2 Confusion matrix

	True class	
Predicted class	True positives (TP)	False positives (FP)
	False negatives (FN)	True negatives (TN)

5. *False Positive Rate*: FPR denotes the proportion of the samples that are not security threats but are incorrectly identified as attacks.

$$FPR = \frac{FP}{FP + TN}$$

6. *False Negative Rate*: FNR denotes the proportion of samples that are security threats but are incorrectly labelled as normal.

$$FNR = \frac{FN}{FN + TP}$$

7. *Negative Predictive Value*: It denotes the total number of true negatives.

$$NPV = \frac{TN}{TN + FN}$$

8. *False Discovery Rate*: FDR denotes the error rate.

$$FDR = \frac{FP}{FP + TP}$$

9. *AUC-ROC Curve*: The Area Under the Curve (AUC) is a metric that measures a classifier's ability to differentiate between classes. The AUC measures how well the model performs at various thresholds between positive and negative classifications. The greater the AUC value, the better. Simply expressed, when AUC equals 1, the classifier can distinguish whether the attack took place or not easily. Figure 4.4 represents AUC-ROC curve for various models used.

It is to be noted that accuracy alone cannot be the only criteria to find whether the model used is the best one or not. Therefore, we need to focus on other metrics as well. Among all these metrics, it is important to ensure that the *False Positive Rate* is minimum, and the detection *Accuracy* is maximum. Table 4.3 provides evaluation metrics for various models used.

It can be observed that all these models give out the same result. Therefore, we can also judge them based on the execution time. Table 4.4 shows the execution time in seconds for each of the models implemented. A comparison of the evaluation time of each classifier is shown in Fig. 4.5.

4.5 Conclusion

In this work, we used various classifiers to detect whether the attack took place or not. Using accuracy and some other metrics as our evaluation criterions, we have evaluated our models. It has been found that except the Random Forest Classifier,

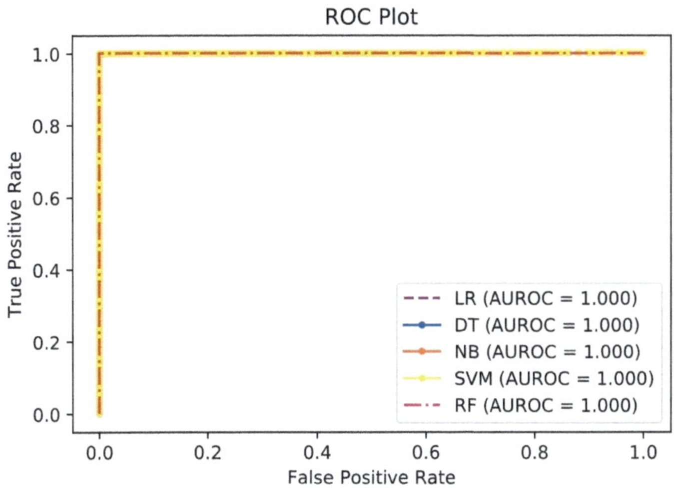

Fig. 4.4 ROC plot

Table 4.3 Evaluation metrics for adults

Classifier	Accuracy	Precision	Recall	F1 score	FPR	FNR	NPV	FDR
LR	100.00	1.00	1.00	1.00	0.00	0.00	1.00	0.00
DT	100.00	1.00	1.00	1.00	0.00	0.00	1.00	0.00
RF	99.99	1.00	1.00	1.00	3.57	0.00	1.00	1.67
NB	100.00	1.00	1.00	1.00	0.00	0.00	1.00	0.00
SVM	100.00	1.00	1.00	1.00	0.00	0.00	1.00	0.00
ANN	100.00	1.00	1.00	1.00	0.00	0.00	1.00	0.00

Table 4.4 Execution time for each classifier

Sr. no.	Classifier	Execution time (s)
1	LR	0.86
2	DT	0.70
3	RF	3.54
4	NB	1.89
5	SVM	10.21
6	ANN	87.33

all other models give a hundred percent accuracy. This is where the execution time comes into place. Among all the models, Decision Tree model takes minimum time and can therefore be considered as an appropriate approach for being used with Intrusion Detection Systems.

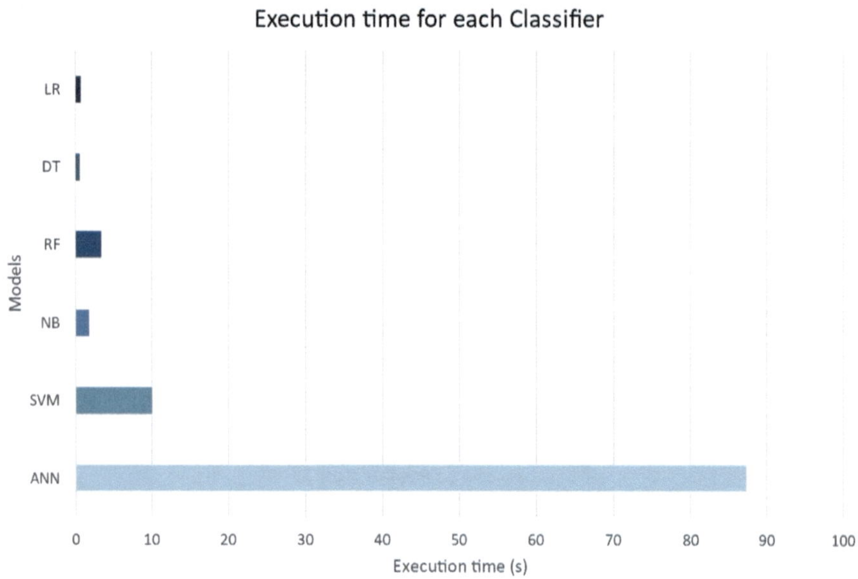

Fig. 4.5 Execution time for each classifier

Finally, it can be concluded that with increasing amount of data in the field of networking, it has become important to device such IDS which can detect potential threats with high accuracy in minimum time.

References

1. N. Moustafa, J. Slay, UNSW-NB15: a comprehensive data set for network intrusion detection systems (UNSW-NB15 network data set), in *2015 Military Communications and Information Systems Conference (MilCIS)* (2015)
2. K. Ee, L.I. Zhichun, G. Jiang, L. Korts-Parn, Z. Wu, Y. Sun, Host level detect mechanism for malicious DNS activities, in *Host Level Detect Mechanism for Malicious DNS Activities. U.S. Patent Appl. 15*, vol. 644 (2018)
3. S.S. Soniya, S.M.C. Vigila, Intrusion detection system: classification and techniques, in *2016 International Conference on Circuit, Power and Computing Technologies (ICCPCT)* (2016)
4. E. Spafford, D. Zamboni, Data collection mechanisms for intrusion detection systems, in *CERIAS Technical Report* (Center for Education and Research in Information Assurance and Security, West Lafayette, 2000), pp. 41315–47907
5. S.M. Kasongo, Y. Sun, Performance analysis of intrusion detection systems using a feature selection method on the UNSW-NB15 dataset. J. Big Data **7**(1) (2020)
6. I. Ahmad, Q.E. Ul Haq, M. Imran, M.O. Alassafi, R.A. AlGhamdi, An efficient network intrusion detection and classification system. Mathematics **10**(3), 530 (2022)
7. D. Jing, H.-B. Chen, SVM based network intrusion detection for the UNSW-NB15 dataset, in *2019 IEEE 13th International Conference on ASIC (ASICON)* (2019)

8. N. Sharma, N.S. Yadav, Ensemble learning based classification of UNSW-NB15 dataset using exploratory data analysis, in *2021 9th International Conference on Reliability, Infocom Technologies and Optimization (Trends and Future Directions) (ICRITO)* (2021)
9. B.S. Bhati, C.S. Rai, B. Balamurugan, F. Al-Turjman, An intrusion detection scheme based on the ensemble of discriminant classifiers. Comput. Electr. Eng. **86**(106742), 106742 (2020)
10. S. Choudhary, N. Kesswani, Analysis of KDD-cup'99, NSL-KDD and UNSW-NB15 datasets using deep learning in IoT. Procedia Comput. Sci. **167**, 1561–1573 (2020)
11. S. Naseer et al., Enhanced network anomaly detection based on deep neural networks. IEEE Access **6**, 48231–48246 (2018)
12. I.H. Sarker, Y.B. Abushark, F. Alsolami, A.I. Khan, IntruDTree: a machine learning based cyber security intrusion detection model. Symmetry (Basel) **12**(5), 754 (2020)
13. J. Ren, J. Guo, W. Qian, H. Yuan, X. Hao, H. Jingjing, Building an effective intrusion detection system by using hybrid data optimization based on machine learning algorithms. Secur. Commun. Netw. **2019**, 1–11 (2019)
14. J. Tarun, et al., Machine learning techniques for prediction of mental health, in *2021 Third International Conference on Inventive Research in Computing Applications (ICIRCA)* (IEEE, 2021)
15. T. Jain, P. Garg, N. Chalil, A. Sinha, V.K. Verma, R. Gupta, SMS spam classification using machine learning techniques, in *2022 12th International Conference on Cloud Computing, Data Science & Engineering (Confluence)* (2022), pp. 273–279. https://doi.org/10.1109/Confluence52989.2022.9734128

Chapter 5
Integration of Machine Learning with Cybersecurity: Applications and Challenges

Suprabha Das, Pranav Gangwani, and Himanshu Upadhyay

Abstract Machine learning (ML) has advanced from the lab to the forefront of operational systems in recent years. Machine learning is used by Facebook, Amazon, and Google every day to improve consumer experiences and purchases. Machine learning enables personal interactions and helps people connect socially with the use of new applications. The significant capability of machine learning is also present in cybersecurity. ML has become necessary in wide range of fields, also there are several cybersecurity implementations of ML. Some of them are malware analysis, particularly for zero-day "malware detection", "threat analysis", "anomaly-based intrusion detection" of typical attacks on sensitive infrastructures, and a variety of other applications.

Keywords Machine learning · Deep learning · Artificial intelligence · Cybersecurity · Malware detection

5.1 Introduction

Machine learning (ML) methods have gained enough fame and acceptance in the world. The current algorithms are getting modified, and the potential to grasp quickly and counter to current problems is valued. As a result of this, ML is now implemented in a majority of fields, including restorative diagnosis, cancer detection [1–3], computer visions, social media marketing, smart transportation [4] and gaming.

S. Das (✉)
Department of Mechanical and Materials Engineering, Florida International University, Miami, USA
e-mail: sdas014@fiu.edu

P. Gangwani
Department of Electrical & Computer Engineering, Florida International University, Miami, USA
e-mail: pgang002@fiu.edu

H. Upadhyay
Applied Research Center, Florida International University, Miami, USA
e-mail: upadhyay@fiu.edu

T. Bhardwaj et al. (eds.), *Artificial Intelligence in Cyber Security: Theories and Applications*, Intelligent Systems Reference Library 240,
https://doi.org/10.1007/978-3-031-28581-3_7

Machine learning techniques, instead of the traditional rule-based learning algorithms or even human interventional approaches, are the better choice in different sectors depending on case by case. Due to these reasons, there is a strong influence in using ML for cyber security. Different detection systems are being modified with machine learning components. Even though the implementation of a completely automated "cyber security" system is still in the future, it is still possible for human personals at the Network and Security Operation Centers (NOC and SOC) to benefit from ML-based prediction and analysis tools.

Web and mobile technologies have evolved rapidly in recent years. Along with this, complexity of the attack strategies for breaching systems and eluding generic signature-based approaches have also increased significantly. Machine learning techniques are capable of adapting quickly to new and unforeseen circumstances. Thus, machine learning can effectively address such challenging and complicated scenarios and provide feasible solutions for the same. Various issues in computer and information security have been addressed using different methods of machine learning.

The two most famous attacks, Malware and ransomware continue to be a big threat to most private, government, and public organizations. Ransomware crippled local governments and healthcare institutions in 2015 by encrypting log files and demanding a "ransom" payment. Ransomware assaults increased by 35% in 2015, according to Symantec [5]. These lucrative assaults enable ransomware creators to sweep millions of dollars in revenue. Ransomware has shifted from conventional IT targets to smart phones and smart gadgets in 2015. Soon, Symantec believes ransomware will spread to IoT wearable devices such as the smart watches and smart homes devices (e.g., thermostats, televisions, etc.).

This chapter is categorized in the following way. Section 5.2 covers an overview of various cyber-attacks, concepts of ML and deep learning (DL) and presents some of the most used datasets related to security. Section 5.3 provides various applications of ML in cybersecurity. Section 5.4 presents the challenges that occur when machine learning is utilized for detection of cyber threats. Finally, Sect. 5.5 concludes the chapter.

5.2 AI-Based Cybersecurity

This section is categorized into four sections. The first part covers the fundamental concepts of machine learning, the second part explains the fundamental concepts of deep learning [6], the third part provides an overview of various cyber threats & attacks, fourth part is on commonly used datasets for providing security in computer systems and mobile applications.

5.2.1 ML and DL Overview

Artificial intelligence (AI) replicates a human brain to identify the best feasible solution to achieve a certain objective. ML is a part of AI that uses results of previous experiences as the basis for future references without being programmed explicitly. There are three primary subcategories of machine learning, "supervised learning", "unsupervised learning", and "semi-supervised leaning".

There is some form of knowledge available about the targeted labels in the data in supervised learning methods. Unsupervised learning is based on recognizing patterns in data with no prior knowledge of the target classes. Semi-supervised learning is a type of ML process that combines both supervised and unsupervised approaches for providing possible solutions. DL is a part of machine learning that has additional capabilities apart from ML techniques. Main distinction between the two is that DL performs actions with several iterations in order to acquire the best potential result. DL strategies tackled the issues from beginning to end, whereas ML techniques divide and conquer. There has been a lot of effort done in the recent decade to employ both strategies to improve cybersecurity. Deep learning can do a job repeatedly without any human contact, whereas ML and DL approaches rely on experience to create input. To create an output, machine learning breaks an issue into smaller pieces, whereas deep learning delivers end-to-end results. For deep learning [7], the time frame of training is higher, but for machine learning, it is shorter. In comparison, the testing period for deep learning [8] is shorter while the testing time for machine learning is more. Deep learning is only possible with a robust hardware system. Based on the dataset which contains low-end information (such as the ones which does require too much space), machine learning works well, However, with DL approaches, they tend to learn from their past experiences and improves overtime based on their failure just like a human brain, whereas ML techniques tend to learn only from previous information of the target "labels".

DL algorithms are divided into two categories: "generative models" and "deep discriminative models". "Recurrent neural networks" (RNN), "Deep Neural Network" (DNN) techniques, and "Convolutional Neural Network" (CNN) are three primary classifications of deep discriminative models. Data is stored in nodes in a recurrent neural network, as the name implies, and all nodes form loops to link to one another. A DNN is one of the most commonly used approach given its major advantages. DNN architecture have several layers. The total number of layers in DNN is supposed to always be greater than three. A CNN on the other hand is a "multi-layer" network that analyzes un-structured data in order to produce complex features as output. Deep Boltzmann machines, and limited Boltzmann machines are the different types of unsupervised/generative models. Two layers are present in a limited Boltzmann machine. A set of weights connects both the hidden and visible levels, but not the same layer. Each layer of a deep belief network acts as a constrained Boltzmann machine. There is one layer called the hidden layer, and the second layer is called a visible layer. A visible layer depicts the data in a deep belief network, whereas a hidden layer depicts the characteristics. The input neurons generated at the output layer is regenerated due to the data loss being reduced by the deep autoencoder, ensuring that the total count of input and output neurons in both layers remains constant. It is a multilayer network with several visible and hidden layers. Each layer in the neuron is connected to its neighbors, but the connections have no directions to each other resulting in no connections between layers.

5.2.2 Cyber Threats and Attacks

Attacking technologies are progressing at a faster pace than defense technologies. Resource protection, data protection, data integrity and data privacy are all goals of cybersecurity. Cyberspace is subject to a variety of dangers and attacks. To mention a few, malware detection, fraud detection, phishing, spam classification, antivirus and disabling firewalls, keystroke recording, malicious URLs, and probing are all common cyber hazards [9].

Phishing and malware are both regarded serious cyber-threats. Phishing is a technique for gaining access to data that is unauthorized by pretending to be a legitimate user. Phishing is when someone sends you a link to a website that looks authentic but takes you to another page where you must provide personal information. However, malware, is harmful software which has been designed specifically to gain unauthorized access to a particular computer and interrupt normal activity flow. Malware detection is divided into three categories: static, dynamic, and hybrid detections. While using static malware detection, apps get inspected for malicious patterns before being executed. Dynamic detection, on the other hand, is done while the programs are executing. Hybrid detection is known to as a combination of the two detection methods. Malware is divided into subcategories like viruses, Trojan horse, and worms. Viruses are malignant parts of software that deletes data from a computer without the knowledge of the user [10]. Worms are malicious software that replicates itself and consumes system resources without authorization.

5.2.3 Common Datasets About Security

When datasets are diverse and real-time data is collected, machine learning approaches provide superior outcomes. The most often used security datasets will be discussed in this section.

The Defense Advanced Research Project Agency (DARPA) datasets, URL dataset, KDD Cup 99 dataset, Australian Defense Force Academy (ADFA) dataset, “HTTP CSIC-2010”, “Android malware dataset”, “Android validation dataset,” “Spambase”, and “NSL-KDD” are some of the most often used security datasets. The identification of attacks is the key objective of the DARPA dataset [11].

Network traffic [12–14] and audit data were used to create the DARPA dataset. When it comes to coping with new system quirks, it has limitations. DARPA fails to display real-world data network traffic. The AFDA dataset was created to outperform the DARPA dataset. In order to accommodate new system changes, AFDA overcame the constraints. A portion of the DARPA dataset was used to create the KDD Cup 99 dataset. The NSL-KDD dataset [15] is a subsequent upgrade for the KDD Cup 99 dataset. To avoid redundancy in data and duplicate instances, "NSL-KDD" was proposed. As a result, the "NSL-KDD dataset" has a fairer amount of records than "KDD Cup 99" and it also gives performance better than "KDD Cup" [16]. The "HTTP CSIC-2010" dataset's primary goal is to detect web attacks. The "HTTP CSIC-2010" dataset processes a large number of web queries. The "HTTP CSIC-2010" dataset is the most well-known and effective dataset for detecting attacks in web searches. The Enron dataset consists of a significant number of emails exchanged by employees of "Enron Corporation". This set of data was used to categorize spam emails. Another widely used dataset for determining and refining spam emails is the Spam base dataset. This dataset computes the various attributes of the gathered observations, and it is available for free on the UCI ML repository. The URL dataset, which is based on internet traffic, was offered as a way to blacklist malicious and harmful URLs. Spam URLs, phishing URLs, malware URLs, defacement URLs and benign URLs are among the five categories of harmful URLs in the URL dataset. The Android malware dataset is based on Android applications. The Android malware dataset [17] was offered as a way to ban malicious Android apps. Two types of characteristics were extracted to build the Android validation dataset: metadata and N-grams, in order to uncover various relationships between 72 genuine apps. The Android validation dataset reveals that applications have various connections, such as siblings, false siblings, stepsiblings, and cousins.

5.3 Applications

Many approaches and techniques for detecting dangers in cyberspace have been developed. Commercial software (antivirus) is useful to identify malware with an excellent success rate, albeit their processes can be disabled by viruses, which evolve and improve at a quicker rate than malware detection tools. Because of the constant growth of threats, learning approaches for detecting unknown malware have been adopted, including a range of machine learning algorithms.

5.3.1 Malware Detection

By researching code patterns and correlations, several machine learning algorithms have been used to detect malicious code in its many forms. In 2005, the first research publications in this field were published, according to the Scopus database and employed an unsupervised machine learning system to investigate computer accesses and Markov models to investigate virus spread. Clustering methods were then used to categorize software as malware, and SVM was tried in both computer systems and Symbian-OS mobile devices [18].

The concept of n-grams, which are continuous sequences of n elements from a text used in natural language processing and computer science, is at the heart of several methodologies. The results are then utilized for classification tasks based on different machine learning algorithms using n-grams to extract features from files or messages.

In 2009, Shabtai et al. presented a state-of-the-art assessment of static feature-based machine learning classifiers, covering decision trees, ANN, Naive Bayes, SVM, boosting approaches, and additional classifiers such the k-nearest neighbor classifier. Later, in one of the significant research papers, naive Bayesian classifiers were used to detect malicious JavaScript code [19]. Later works, such as [20, 21], successfully applied the Naive Bayes model.

There have been substantial recent efforts in malware detection, and the SVM technique has been frequently employed with positive results. Biggio et al. [22] carried out a detailed analysis on SVM for security applications like malware detection, finding the three most dangerous acts that the classification algorithm may face: poisoning (deceiving the system), evasion (avoiding detection), and privacy breaches (obtaining information from internal variables). Figure 5.1 shows how a machine learning model works for detecting malware in data.

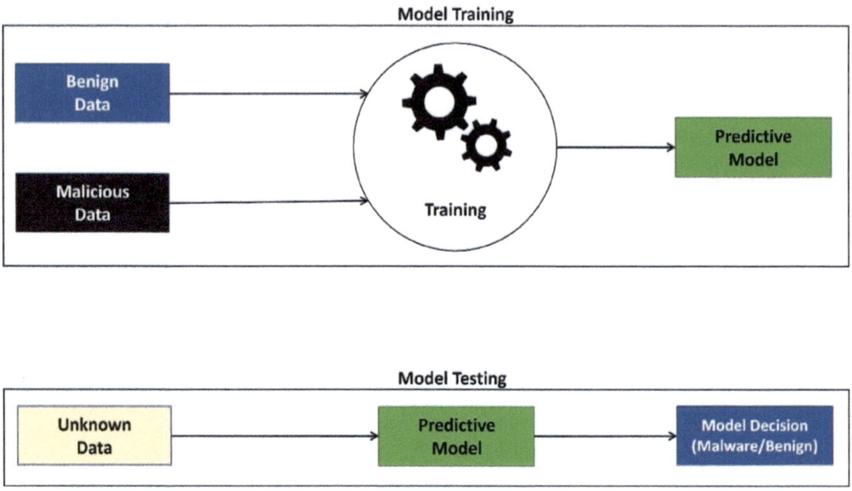

Fig. 5.1 Malware detection training and testing

5.3.2 Power System Security

Electricity supply networks are critical to the economy and security of every country. Extreme forms of power loss are called blackouts [23], that affect a large area with significant societal effects, can be caused by a variety of factors including a power transmission line accident or deliberate threats. Initial failures propagate in a diversified and intricate cascade of unusual events, resulting in widespread blackouts. When the electrical system is strained, however, the severity of the outage may be exacerbated by additional line trips, contribute to the disturbance's geographical transmission, and even result in a cascading blackout. As a result, a small cyber-attack during a stressed power source can be disastrous due to catastrophic failures.

Numerous new digital technologies are increasing the surfaces for cyber-attacks in power grids, such as digital protection relays in power systems [24] adaptive protection approaches, "phasor measurement units (PMUs)", and so on. The authors employed a machine learning technique for detecting the strained state of the power system in "(Bernabeu et al. 2012)". They built a "discrimination function" that categorizes the system state as stressed or safe using a decision tree-based technique. PMU's potential predictors, such as "voltage magnitudes", "angle differences", "mega volt-amps flows", "current magnitudes", and so on, are used to train the classifier.

In order to forecast the necessary dependability balance of the adapted protective scheme based on large area measurements, the supervised ML algorithm decision trees (DT) is utilized. ML has also been utilized to enhance the quantity of PMUs to be placed at their most critical position. The decision tree aids in the division of features that decide where "PMUs" should be employed. In spite of the increasing uncertainty in power systems, ML aids engineers in the power-system, in meeting the problem of designing and managing future power systems while maintaining the security level.

5.3.3 Phishing Detection

When a record number of phishing attacks were recorded in January 2006, [25] it was extremely noteworthy. Before that, there were a few studies on phishing, but in 2007, two major research papers focusing on detecting phishing e-mails using machine learning approaches were published: PILFER [26], a machine-learning-based classification technique used the sci-kit learn algorithm called decision tree and achieved excellent results. The second study compared a variety of ML algorithms, including SVM, ANN, random forest, logistic regression, with mixed results: the random forest algorithms produced the highest accuracy, while logistic regression had the lowest false positive rate and weighted error rate. Figure 5.2 shows the various applications of machine learning in the field of cybersecurity.

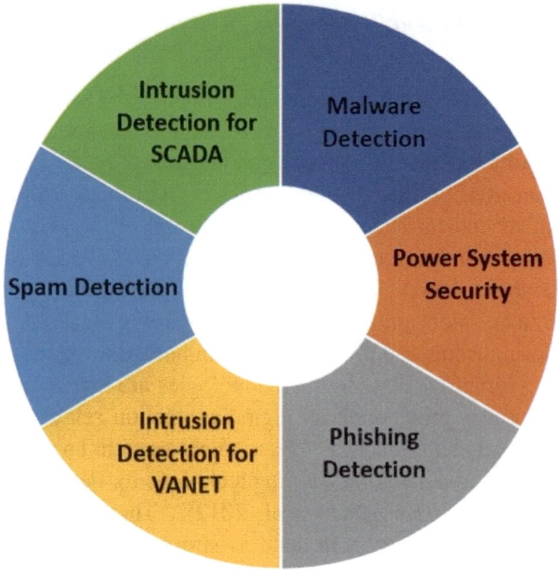

Fig. 5.2 Applications of machine learning in cybersecurity

Since then, the SVM approach has been used to classify phishing websites in [27], identify phishing URLs in [28], and detect phishing e-mails.

5.3.4 Intrusion Detection in SCADA Systems

SCADA systems are important for controlling and monitoring such as electric power production, distribution and transmission, or a water/sewage treatment facility. However, because SCADA systems are increasingly connected to IT networks, their security [29] has grown more challenging. This was done in order to improve the connection between field operations and the business network. Hackers' dangers and hazards to the SCADA system have escalated as a result. As a result, there is a strong need to create and develop certain alarm systems that can assist CI operators in identifying continuing intrusions. These systems must be better protected against cyber-attacks like the SLAMMER worm [30]. Two US utilities as well as a nuclear power plant were infected by worm, resulting in a blackout in the US's northeast. As a result, "intrusion detection systems (IDS)" must be developed, which are utilized to detect a threat and send out correct notifications that can assist in taking necessary action. It is possible that IDS is not able to handle all types of threats and may create false alerts. These false alerts might cause high economic risk. To distinguish between regular and malicious communications, an analyst can use machine learning [31] methods such as support vector machine (SVM). SVM is a data classification method. One-Class SVM (OCSVM) is a supervised ML approach for detecting new or unfamiliar data that produces a decision function.

Two distributed IDS that can detect assaults in a SCADA system were proposed in [32]. It is based on real-time perimeter IDS, which provides basic cyber-analysis for evaluating and safeguarding the security perimeter of each CI. Using datasets from a small-scale test bed, two OCSVM modules were built and evaluated. The first technique is "K-OCSVM" [33], which assists in the distinction between true and false alerts. It's a hybrid of the OCSVM and the K-means clustering methods "(Kanungo et al. 2002)". "K-OCSVM" is distinct from other comparable approaches that employ cross-validation to determine parameters. Cross-validation is a statistical approach for testing how well a statistical study's conclusions will generalize to a new collection of data [34, 35]. The detection module's output is sent to the system via "intrusion detection message detection exchange format (IDMEF)" files, which contain information about the intrusion's source, timing, and severity. "K-OCSVM" is trained offline using network traces, and the detection module's output is sent to the system via intrusion detection message detection exchange format (IDMEF) files, which contains data about the intrusion's source, timing, and gravity. This method has a good performance in terms of low overhead and accuracy, however, it does so by not considering minor changes in the communication network which might obscure attacks.

IT-OCSVM is one more method that was presented in [36, 37] which produced low overhead and high accuracy. The architecture of IT-OCSVM is displayed in Fig. 5.3 and contains six steps. In the first step, the raw data acquired from the testbed is preprocessed. In the second step, feature selection is performed which includes features that depend on time and content. In the third step, the entire data is split, and each single dataset is then trained using OCSVM. The central OCSVM runs concurrently with the cluster of divided OCSVMs. In the fourth step, the models are tested with a new dataset after the training is completed. In step five, an ensemble model is used which combines the results of various OCSVM modules. In the last step, all the results of various models are collected and analyzed.

5.3.5 Spam Detection

Spam is frequently connected with inbound e-mails, but it has also been identified on blogs search engines, and even tweets. Spam is identified using filters that evaluate the content to identify whether it is fake spam or genuine website or email.

Spammers easily avoided the user-defined first filters by exploiting content "obfuscation". Many ML algorithms have been used and developed for Spam detection in the recent years.

The following criteria must be addressed by any system targeted at identifying spam: (1) shifting class distributions, (2) message misclassification costs, (3) complicated text patterns, (4) changing target, and (5) smart adaptive opponents. To detect spam, there are two basic strategies: (1) textual analysis and (2) image-based analysis. The initial approach was textual analysis, which is essentially a text classification

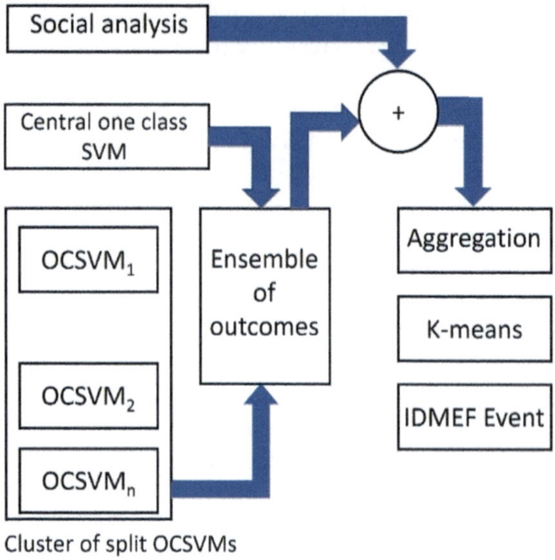

Fig. 5.3 Working of IT-OCSVM

issue. As spam communications became more complex and included graphics with embedded text, image-based analysis became essential.

Filters follow a similar pattern: Initially, message's essential words are extracted (tokens) and reduce things to their most fundamental forms; next, they eliminate frequent and unnecessary terms, and the classifier receives the generated set of words as input. Next paragraphs shall go over the various classifiers.

Bayesian classifiers usually are employed and consider the likelihood of knowing if a message is a spam message. These classifiers often use a "bag-of-words (BoW) model", which is a basic representation of words used in communications that is commonly employed in document categorization.

5.3.6 Intrusion Detection Systems for VANETs

VANET [38] is a new methodology that provides safety and vital information in current transportation networks. It offers the advantages of improved road safeness and traveler comfort while also safeguarding the privacy of the driver from various forms of attacks. They are typically subject to a variety of passive and active threats, including eavesdropping and interference. By identifying aberrant or malicious actions, IDS can be used to minimize threats for instance, regulation breaches and illegitimate invasions. If vehicles in the VANET collaborate, this detection can be done more accurately. A VANET-based collaborative detection technique may be developed using distributed machine learning.

However, one of the most serious problems with "collaborative learning" is that nodes share information with one another, putting consumer privacy at danger. A malicious node can cause havoc and steal important information from other nodes. "Zhang and Zhu" [39] have presented a "machine-learning-based collaborative IDS (CIDS)" architecture (2018). In essence, it's used to train a classifier to detect VANET intrusions. The "CIDS" allows cars to utilize the labeled data of other vehicles for training. It effectively expands the training data for each vehicle. As a consequence, because the workload is spread across all of the vehicles in the system, the burden on each vehicle is decreased. Vehicles can also communicate data without sharing training data thanks to CIDS.

The most important danger is a breach of privacy, as any malicious outsider can see their categorization results. It's a well-defined notion that ensures that every modification to a single dataset item has only a little impact on the distribution of the dataset's responses, providing a strong privacy guarantee. As a result, this strategy ensures road safety while also protecting the driver's information.

5.4 Challenges

In this section, we present and describe various challenges that arise when machine learning in integrated with the cybersecurity domain.

5.4.1 Lack of Labeled Samples and Certainty in Ground Truth

The majority of cyber-security tasks are supervised learning exercises. We must determine whether an entity or action is malevolent or benign. However, we frequently lack the labels necessary for supervised learning [40].

The breadth of manual labeling is limited and subjective. As a result, the majority of labels are based on heuristics. Apart from making mistakes, one can end up attempting to imitate the heuristics with which he began.

A popular strategy is to rely on local expertise. When the entity to classify is universal (occurring in cases other than the one under investigation, such as malware or a malicious domain but not network activity), external black and whitelists might be used as labels. Because the lists are not from the same source as the things to classify, this method may cause a domain adaptation difficulty. Naturally, such listings may be of poor quality.

Another popular solution is a consensus agreement, which is based on the judgment of numerous vendors. This strategy has algorithmic issues in addition to potential intellectual property issues. The ground truth does not always align with the consensus. Advanced Persistent Threats (APT) are a good example of this dilemma.

APT attacks are typically sophisticated and well-targeted. Because an APT would often utilize its software on a single target, threat intelligence will not recognize it, and the virus will elude detection. Using consensus agreement as a concept is like to instructing your model to ignore APT detection.

Unsupervised learning approaches, notably anomaly detection, may be used in algorithmic solutions. While anomaly detection is widely utilized in many fields, it is rarely used in cyber-security due to intrinsic problems, which we refer to as machine learning challenges. Because some domains, such as networks, have a variety of benign abnormalities, these solutions may be troublesome. A sophisticated attacker will most likely be aware of the defense and will attempt to make its activities appear normal. We may wind up with a slew of oddities in this circumstance, none of which are related to the attacker.

5.4.2 Imbalanced Data Sets

When the ratios between the majority and minority sets are significant, we consider a data set to be imbalanced. While a ratio of 1 to 10 is considered imbalanced in the machine learning community, harmful training samples in cybersecurity data sets [41] may be uncommon, and imbalance ratios of 1 to 10,000 are frequent.

The term relative imbalance is used when the ratio between the sets is considerable, but each set has adequate samples. This is total imbalance in the more severe case where there aren't enough samples in the minority set. This is a typical occurrence in the field of cyber-security.

5.4.3 Concept Drift

A security expert successfully overcomes the aforementioned hurdles and launches a high-performing system to the production environment. The system's performance degrades dramatically in a short period of time. The model remained unchanged. It was the planet itself that had transformed. Cybersecurity is a field that is changing fast. Models and insights are soon becoming obsolete. As technology advances, new protocols and areas emerge, and malevolent activity develops its own trends and fads. Concept drift [42] refers to a shift in the source of a studied thing as time passes. To avoid having to restart from the beginning, methods for tracking idea drift and radical changes in the domain must be established.

5.4.4 *Domain Adaptation*

A case in which the test distribution used to evaluate a model differs from the train distribution used to generate the model is known as domain adaptation. In each of its sections, the cyber world appears differently. Parties engaging at distinct scales, domains, cultures, and geo-locations have varied threats, entities, and networking behavior. Adapting an effective defense from one domain to another is thus a fundamental challenge.

5.5 Conclusion

In this chapter, we comprehensively reviewed how machine learning can be integrated with cybersecurity. Machine learning when integrated with cybersecurity can provide more security to various systems and detect malicious activities and vulnerabilities real-time. We presented many applications of machine learning and cybersecurity that are currently employed in various industries. Furthermore, we presented various challenges that arise when machine learning is integrated with cybersecurity. From our study, the main challenge is acquiring enough data for the machine learning models to detect fraudulent activities.

References

1. T. Saba, Recent advancement in cancer detection using machine learning: systematic survey of decades, comparisons and challenges. J. Infect. Public Health **13**(9), 1274–1289 (2020). https://doi.org/10.1016/j.jiph.2020.06.033
2. K. Bhanot, S.K. Peddoju, T. Bhardwaj, A model to find optimal percentage of training and testing data for efficient ECG analysis using neural network. Int. J. Syst. Assur. Eng. Manag. **9**(1), 12–17 (2018). https://doi.org/10.1007/s13198-015-0398-7
3. K. Kadarla, S. C. Sharma, T. Bhardwaj, A. Chaudhary, A simulation study of response times in cloud environment for IoT-based healthcare workloads, in *Proceedings of the 14th IEEE International Conference on Mobile Ad Hoc Sensor Systems MASS 2017*, (2017), pp. 678–683. https://doi.org/10.1109/MASS.2017.65
4. D. Gangwani, P. Gangwani, Applications of machine learning and artificial intelligence in intelligent transportation system: a review, in *Lecture Notes in Electrical Engineering* (Springer, 2021), pp. 203–216
5. Symantec, Internet security threat report. Netw. Secur. **21**(2), 1–3 (2016)
6. T. Bhardwaj, R. Mittal, H. Upadhyay, L. Lagos, Applications of swarm intelligent and deep learning algorithms for image-based cancer recognition, in *Artificial Intelligence in Healthcare* (Springer, Singapore, 2022), pp. 133–150
7. P. Gangwani, J. Soni, H. Upadhyay, S. Joshi, A deep learning approach for modeling of geothermal energy prediction. Int. J. Comput. Sci. Inf. Secur. **18**(1), 62–65 (2020)
8. T. Bhardwaj, H. Upadhyay, L. Lagos, Deep learning-based cyber security solutions for smart-city: application and review," in *Artificial Intelligence in Industrial Applications*, vol. 25, ed by T. Sharma, S. Fernandes (Springer, Cham, 2022)

9. T. Bhardwaj, T. K. Sharma, M. R. Pandit, Social engineering prevention by detecting malicious URLs using artificial bee colony algorithm. Adv. Intell. Syst. Comput. **258**, 355–363 (2014). https://doi.org/10.1007/978-81-322-1771-8_31
10. T. Bhardwaj, End-to-End Data Security for Multi-Tenant Cloud Environment (2016)
11. M.M. Anjum, S. Iqbal, B. Hamelin, Analyzing the usefulness of the DARPA OpTC dataset in cyber threat detection research, in *Proceedings of the 26th ACM Symposium on Access Control Models and Technologies* (2021), pp. 27–32. https://doi.org/10.1145/3450569.3463573
12. T. Bhardwaj, C. Reyes, H. Upadhyay, S.C. Sharma, L. Lagos, Cloudlet-enabled wireless body area networks (WBANs): a systematic review, architecture, and research directions for QoS improvement. Int. J. Syst. Assur. Eng. Manag. (2021). https://doi.org/10.1007/s13198-021-01508-x
13. T. Bhardwaj, S.C. Sharma, Fuzzy logic-based elasticity controller for autonomic resource provisioning in parallel scientific applications: A cloud computing perspective. Comput. Electr. Eng. **70**, 1049–1073 (2018). https://doi.org/10.1016/j.compeleceng.2018.02.050
14. T. Bhardwaj, S.C. Sharma, Cloud-WBAN: an experimental framework for cloud-enabled wireless body area network with efficient virtual resource utilization. Sustain. Comput. Informatics Syst. **20**, 14–33 (2018). https://doi.org/10.1016/j.suscom.2018.08.008
15. B. Ingre, A. Yadav, Performance analysis of NSL-KDD dataset using ANN, in *2015 International Conference on Signal Processing and Communication Engineering Systems* (2015), pp. 92–96. https://doi.org/10.1109/SPACES.2015.7058223
16. M. Tavallaee, E. Bagheri, W. Lu, A. A. Ghorbani, A detailed analysis of the KDD CUP 99 data set, in *2009 IEEE Symposium on Computational Intelligence for Security and Defense Applications* (2009), pp. 1–6. https://doi.org/10.1109/CISDA.2009.5356528
17. Y. Zhou, X. Jiang, Dissecting android malware: characterization and evolution, in *2012 IEEE Symposium on Security and Privacy* (2012), pp. 95–109. https://doi.org/10.1109/SP.2012.16
18. A.-D. Schmidt, J.H. Clausen, A. Camtepe, S. Albayrak, Detecting Symbian OS malware through static function call analysis, in *2009 4th International Conference on Malicious and Unwanted Software (MALWARE)* (2009), pp. 15–22. https://doi.org/10.1109/MALWARE.2009.5403024
19. Y. Hao, H. Liang, D. Zhang, Q. Zhao, B. Cui, JavaScript malicious codes analysis based on naive bayes classification, in *2014 Ninth International Conference on P2P, Parallel, Grid, Cloud and Internet Computing* (2014), pp. 513–519. https://doi.org/10.1109/3PGCIC.2014.147
20. Y. Lu, P. Zulie, L. Jingju, S. Yi, Android malware detection technology based on improved Bayesian classification, in *2013 Third International Conference on Instrumentation, Measurement, Computer, Communication and Control* (2013), pp. 1338–1341. https://doi.org/10.1109/IMCCC.2013.297
21. F. Shang, Y. Li, X. Deng, D. He, Android malware detection method based on naive Bayes and permission correlation algorithm. Cluster Comput. **21**(1), 955–966 (2018). https://doi.org/10.1007/s10586-017-0981-6
22. B. Biggio et al., Security evaluation of support vector machines in adversarial environments. Support Vector Mach. Appl. **9783319023007**, 105–153 (2014). https://doi.org/10.1007/978-3-319-02300-7_4
23. H. Haes Alhelou, M. Hamedani-Golshan, T. Njenda, P. Siano, A survey on power system blackout and cascading events: research motivations and challenges. Energies **12**(4), 682 (2019). https://doi.org/10.3390/en12040682
24. M. Kezunovic et al., Design, implementation and validation of a real-time digital simulator for protection relay testing. IEEE Trans. Power Deliv. **11**(1), 158–164 (1996). https://doi.org/10.1109/61.484012
25. Z. Ramzan, C. Wüest, Phishing attacks: analyzing trends in 2006, in *4th Conference on Email Anti-Spam, CEAS 2007* (2007)
26. S.O. Uwagbole, W.J. Buchanan, L. Fan, Applied machine learning predictive analytics to SQL injection attack detection and prevention, in *2017 IFIP/IEEE Symposium on Integrated Network and Service Management (IM)* (2017), pp. 1087–1090. https://doi.org/10.23919/INM.2017.7987433

27. A. Altaher, Phishing websites classification using hybrid SVM and KNN approach. Int. J. Adv. Comput. Sci. Appl. **8**(6) (2017). https://doi.org/10.14569/ijacsa.2017.080611
28. M. Zouina, B. Outtaj, A novel lightweight URL phishing detection system using SVM and similarity index. Human-centric Comput. Inf. Sci. **7**(1), 17 (2017). https://doi.org/10.1186/s13673-017-0098-1
29. P. Gangwani, A. Perez-Pons, T. Bhardwaj, H. Upadhyay, S. Joshi, L. Lagos, Securing environmental IoT data using masked authentication messaging protocol in a DAG-based blockchain: IOTA tangle. Futur. Internet **13**(12), 312 (2021). https://doi.org/10.3390/fi13120312
30. D. Moore, V. Paxson, S. Savage, C. Shannon, S. Staniford, N. Weaver, Inside the slammer worm. IEEE Secur. Priv. **1**(4), 33–39 (2003). https://doi.org/10.1109/MSECP.2003.1219056
31. D. Gangwani, Q. Liang, S. Wang, X. Zhu, An empirical study of deep learning frameworks for melanoma cancer detection using transfer learning and data augmentation, in *2021 IEEE International Conference on Big Knowledge (ICBK)* (2021), pp. 38–45. https://doi.org/10.1109/ICKG52313.2021.00015
32. W. Gao, T. Morris, B. Reaves, D. Richey, On SCADA control system command and response injection and intrusion detection, in *2010 eCrime Researchers Summit* (2010), pp. 1–9. https://doi.org/10.1109/ecrime.2010.5706699
33. L.A. Maglaras, J. Jiang, OCSVM model combined with K-means recursive clustering for intrusion detection in SCADA systems, in *10th International Conference on Heterogeneous Networking for Quality, Reliability, Security and Robustness* (2014), pp. 133–134. https://doi.org/10.1109/QSHINE.2014.6928673
34. T. Bhardwaj, S.C. Sharma, An autonomic resource provisioning framework for efficient data collection in cloudlet-enabled wireless body area networks: a fuzzy-based proactive approach. Soft Comput. **23**(20), 10361–10383 (2019). https://doi.org/10.1007/s00500-018-3587-x
35. R. Panwar, M. Supriya, Autonomic resource allocation frameworks for service-based cloud applications: a survey, in *Proceedings of the 2019 International Conference on Computing, Communication and Intelligent Systems ICCCIS 2019*, vol. 2019 (2019), pp. 214–219. https://doi.org/10.1109/ICCCIS48478.2019.8974463
36. L.A. Maglaras, J. Jiang, T.J. Cruz, Combining ensemble methods and social network metrics for improving accuracy of OCSVM on intrusion detection in SCADA systems. J. Inf. Secur. Appl. **30**, 15–26 (2016). https://doi.org/10.1016/j.jisa.2016.04.002
37. S. Shaw, S. Kadam, S. Joshi, D. Hadsul, *Advanced Virtual Apparel Try Using Augmented Reality (AVATAR)*, vol. 1154 (2020)
38. Y. Tang, N. Cheng, W. Wu, M. Wang, Y. Dai, X. Shen, Delay-minimization routing for heterogeneous VANETs with machine learning based mobility prediction. IEEE Trans. Veh. Technol. **68**(4), 3967–3979 (2019). https://doi.org/10.1109/TVT.2019.2899627
39. T. Zhang, Q. Zhu, Distributed privacy-preserving collaborative intrusion detection systems for VANETs. SIEEE Trans. Signal Inf. Process. over Networks **4**(1), 148–161 (2018). https://doi.org/10.1109/TSIPN.2018.2801622
40. K. Shaukat et al., Performance comparison and current challenges of using machine learning techniques in cybersecurity. Energies **13**(10), 2509 (2020). https://doi.org/10.3390/en13102509
41. M. Pawlicki, M. Choraś, R. Kozik, W. Hołubowicz, On the impact of network data balancing in cybersecurity applications, in *Lecture Notes in Computer Science* (2020), pp. 196–210
42. S. Singhal, U. Chawla, R. Shorey, Machine learning & concept drift based approach for malicious website detection, in *2020 International Conference on COMmunication Systems & NETworkS (COMSNETS)* (2020), pp. 582–585. https://doi.org/10.1109/COMSNETS48256.2020.9027485

Chapter 6
AI-Based Anomaly Detection on IoT Data-Driven Thermal Power Plants for Condition Monitoring and Preventive Maintenance

Pranav Gangwani, Santosh Joshi, Himanshu Upadhyay, and Leonel Lagos

Abstract Anomalies are an integral part of every system's behavior and sometimes cannot be avoided. The thermal power plant systems are one of the most complex dynamical systems which must function properly all the time with the least amount of costs. Therefore, it is crucial to timely detect such anomalies in real-world running power plant systems for monitoring the condition of a component and to avoid any failure or future maintenance. This book chapter discusses the need for detecting anomalies in thermal power plant data for monitoring and maintaining the conditions of these power stations and discusses various Artificial Intelligence (AI) techniques to detect sensor data anomalies in different components of a thermal power plant.

Keywords IoT · Sensors · Machine learning · Deep learning · Artificial intelligence · Anomaly detection · Thermal power plant

6.1 Introduction

For accurate analysis of thermodynamic systems, a significant number of assumptions are needed to be made to account for numerous uncertainties in the solution. When idealistic assumptions like these are not feasible in practical applications, analyzing these systems necessitates solving thousands of nonlinear equations,

P. Gangwani (✉)
Department of Electrical and Computer Engineering, Florida International University, Miami, USA
e-mail: pgang002@fiu.edu

S. Joshi · H. Upadhyay · L. Lagos
Applied Research Center, Florida International University, Miami, USA
e-mail: sajoshi@fiu.edu

H. Upadhyay
e-mail: upadhyay@fiu.edu

L. Lagos
e-mail: lagosl@fiu.edu

T. Bhardwaj et al. (eds.), *Artificial Intelligence in Cyber Security: Theories and Applications*, Intelligent Systems Reference Library 240,
https://doi.org/10.1007/978-3-031-28581-3_8

resulting in high computational efforts. Machine/deep learning algorithms [1] are receiving more attention to overcome these constraints by bypassing methods based on thermodynamics [2], identifying non-intuitive characteristics, and providing estimated performance beyond the model parameters. Studying the various complicated correlations and interdependencies between significant input and output factors has resulted in such advancements.

Full load plant performance, often known as maximum power production, is an excellent example of complicated thermodynamic systems [3] and dependent on multiple factors for instance pressure of the vacuum, condition of weather, procedures that are interconnected, and so on, making accurate physics-based models tough to develop. However, accurately estimating full load power [4] production is critical for efficient and profitable optimizing plant efficiency and profitability. Understanding various complicated correlations and interdependencies between key input and output factors is the source of such advancements.

Other important factors that have a significant impact on power plant performance, contractual duty, and income is the unexpected downtime or system and sensor [5] breakdown. Downtime has become an immensely important problem since the capabilities of design of thermal power stations have dramatically increased for the past several years [6]. Even for a small 560 MW power plant today, unpredictable interruptions can result in significant losses. Simultaneously, faulty equipment endangers not just the station's profitability but also its security. As a result, performance estimate and health monitoring are two critical responsibilities in power plant operations and maintenance.

Anomalies are data patterns [7] that do not adhere to a well-known idea of normal behavior in data mining. The challenge of differentiating anomalies and typical values from normal or anticipated behavior is referred to as anomaly detection, sometimes known as novelty detection. There are various sorts of defects that can arise in the functioning of a power plant, and these problems occur randomly. In this case, gathering adequate representatives for all sorts of defects from on-site observations is too costly, rendering classification techniques ineffective for condition monitoring. The goal of anomaly detection [8] in conditioning monitoring is to detect the development of faults at an earlier stage and provide assistance in human decision-making.

6.2 Condition Monitoring and Preventive Maintenance

6.2.1 Condition Monitoring

The monitoring of structural integrity, as well as the performance of various processes of the major components of a thermal power plant, is known as plant condition monitoring. The use of a combination of online Internet of Things (IoT) devices [9] for condition monitoring and instruments (both traditional and new) are some of the necessities for this task [10]. Some more requirements include laboratory testing, offline methods for non-destructive testing, and various tools for modeling. In terms of structural integrity evaluation, the offline NDT methods concerning the lower diagnostic reliability and high costs of online methods of monitoring are more than compensated for by their capacity to provide time-continuous structural risk reduction. The employed combination of offline and online methods can achieve the best compromise on cost–benefit, and this can only be ensured by problem-specific operational experience.

6.2.2 *Preventive Maintenance*

The main causes of degraded reliability and worsened performance are failures that occurred onsite in the operation phase and these failures [11] must be avoided to assure reliability. Proper maintenance is the best way to reduce possible failures in the operation of a thermal power plant. Mainly, there exist three categories of maintenance: corrective maintenance (CM), preventive maintenance (PM), and improvement maintenance (IM) [12]. IM is done at the time of designing and focuses on complete failure elimination. The aim of IM is the reduction or total elimination of the need for maintenance. A designer, however, must deal with a variety of issues and limitations such as budget, utilization of space, and requirements of the market and the cost of a product is dependent on its reliability. However, the repair or maintenance which is done after the failures have occurred is known as CM [13]. The third category of maintenance is PM which is done to avoid failures and keep the equipment in a good operating system. The job of PM is to alert when a failure is about to happen and accordingly, maintenance can be done before some significant damage is caused by the failures or a loss of major capital has occurred. Therefore, for promoting reliability and durability, PM is the best approach. PM consists of two main approaches which are called condition-based maintenance and time-based maintenance. The main goal of PM no matter the approach adopted is the early detection of a failure or even predict a potential failure if possible. If the variables or the parameters in the device data indicate that a failure is about to happen then that failure can then be avoided.

6.3 AI-Based Anomaly Detection

Earlier, for anomaly detection in thermal power plants supervised machine learning algorithms [14] were widely studied and adopted. For prognosis and fault diagnosis in various thermal power plant components such as combustors and gas turbine engines, supervised neural networks [15] have been the most popular techniques. Recently, for anomaly detection [16] in thermal power plants and other fields, the use of unsupervised learning algorithms [17, 18] is increasing rapidly. The main benefit of using unsupervised learning techniques is that they do not need training instances that are labeled [19], unlike supervised algorithms [20]. This can become a great contribution for detecting anomalies [21] in thermal power plants at an early stage since acquiring enough labeled data [22] can be hard. In this section, we identify and explain the working of various machine learning [23] and deep learning algorithms [24] for anomaly detection [25] which have been used or currently employed to detect anomalies in thermal power plants.

6.3.1 Machine Learning-Based Anomaly Detection

(1) One-Class Support Vector Machine (SVM)

The working of Support Vector Machine (SVM) [26] is based on a simple idea as described in this section. A nonlinearly separable data is chosen which can be transformed into a high dimensional space. This is done by using a map that is nonlinear to separate the data point linearly. Our approach defines the data about the attacks as negative and the normal data is defined as the positive example in our case. In [27] a theory was mentioned on SVM which uses a 2-class SVM [28] to separate the hyper-plane according to the support vectors placed adjacent to the hyper-plane. However, if the total number of negative instances when is less, the SVM classifier's performance in terms of generalization becomes poor. Moreover, about the facts mentioned above, the error rates are unsatisfactory.

The theory of SVM was applied to the one-class classification problem by proposing a method in [27]. In general, they viewed the origin after changing the features [29] with a kernel, as the only member of the second class. They removed the picture of one class from the origin by using a "relaxation parameter".

The main notion with the feature space is first, presuming that origin is in the second class and all the instance points are adjacent to the origin and that is to be contemplated as the data points or outliers [30] with anomalies. Once the input data and the samples chosen by the method match, they will be considered as anomaly data and hence belongs to the anomalous category. The geometric interpretation is depicted by Fig. 6.1 which shows the one-class SVM.

In such cases, determining how far away a point can be from the origin before it is considered to be anomalous data is advisable.

Based on the training data given to us $\{(x_1, y_1), (x_2, y_2), \ldots, (x_l, y_l)\}$, where $x \in R^n$, $y \in \{-1, +1\}$, and R^n is considered as the feature space, we can come up with the following quadratic programming problem:

$$\begin{aligned} &\min\left(R^2 + \mathrm{C}\sum_{\mathrm{i}=1}^{1} \varepsilon_{\mathrm{i}}\right) \\ &s.t.\ y_i(\|\emptyset(x_i) - a\|^2 - R^2) \le \varepsilon_i \\ &\varepsilon_i \ge 0, 1 \le i \le l \end{aligned} \tag{6.1}$$

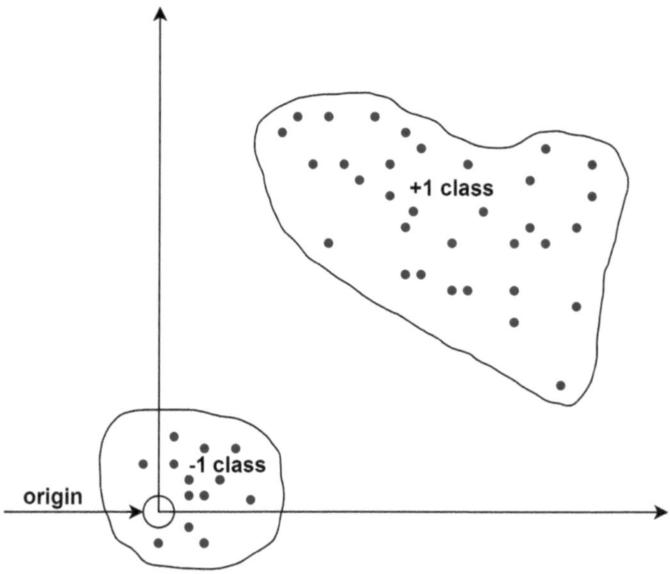

Fig. 6.1 Geometrical illustration of one-class SVM model

where ε_i stands for the slack variables which are reprimanded in the objective function. The purpose of implementing a slack variable is to permit for some amount of accuracy during the training, where a is the center and R is the radius of hyper-sphere, and C is considered as the penalty parameter. The Lagrangian form corresponding to the Eq. (6.1) is as follows:

$$\begin{aligned} L = R^2 + \mathrm{C}\sum_{\mathrm{i}=1}^{1} \varepsilon_{\mathrm{i}} \\ + \sum_{i=1}^{l} \alpha_i (y_i (\|\emptyset(x_i) - a\|^2 - R^2) - \varepsilon_i) \\ - \sum_{i=1}^{l} \beta_i \varepsilon_i \end{aligned} \tag{6.2}$$

where $\alpha_i \geq \beta_i \geq 0$ are considered as Lagrange multipliers. This formula is generated by differentiating with respect to all the variables to zero:

$$\sum_{i=1}^{l} \alpha_i y_i = 1,$$
$$0 \leq \alpha_i \leq C,$$

$$\sum_{i=1}^{l} \alpha_i y_i \emptyset(x_i) = a \tag{6.3}$$

the corresponding dual is:

$$\min\left(\sum_{i,j=1}^{l} \alpha_i \alpha_j y_i y_j (\emptyset\emptyset(x_i).\emptyset x_j)\right) - \sum_{i=1}^{l} \alpha_i y_i (\emptyset(x_i).\emptyset(x_i)))$$
$$s.t.\ 0 \le \alpha_i \le \mathrm{C}, \sum_{\mathrm{i}=1}^{1} \alpha_{\mathrm{i}} \mathrm{y}_{\mathrm{i}} = 1 \tag{6.4}$$

We use some suitable kernel function $K(x_i, x_j)$ to represent the inner product $\varnothing(x_i).\varnothing(x_j)$. To make the right decision of choosing a kernel function depends on the experience of training a model and experiment results. We apply a Gaussian RBF $K(x, y) = e^{-\|x-y\|/q}$ in this method.

After this, the center of the hyper-sphere and the distance among the instance points for any input x is calculated, only when the below requirement is fulfilled

$$||\phi(x) - x|| \le R \tag{6.5}$$

the data point x lies in the hyper-sphere and the +1 class [31], or else it fits the −1 class. The radius of the hyper-sphere can be reckoned with the distance between the bounded and the center of the hyper-sphere by the following method:

$$R^2 = 1 - \frac{2}{n_{msv}} \sum_k \sum_i \alpha_i y_i K(x_k, x_i) + \sum_{i,j} \alpha_i \alpha_j, y_i y_j K(x_i x_j) \tag{6.6}$$

where x_k are known as the bounded vectors and n_{msv} is considered as the number of bounded vectors in the equation. Hence, the decision can be shown as a function given below:

$$f(x) = sgn\left(\sum_{i=1}^{l} \alpha_i y_i K(x, x_i) + b\right) \tag{6.7}$$

where, $b = -\frac{1}{n_{msv}} \sum_k \sum_i \alpha_i y_i K(x_k, x_j)$. For a given vector x, someone can determine if it lies in class +1, if $f(x) \ge 0$ otherwise, it belongs to class −1. While x does not belong to class −1 or $y_j = +1$, which is similar to the one-class SVMs.

(2) Isolation Forest

There are two anomaly data features [32] on which the Isolation Forest algorithm [33] is built: a. the exception data makes up a lesser percentage of the whole data set size, b. feature value of the normal data [34] and abnormal data are significantly different. The data [35] is recursively separated in a training set with only the continuous category until the isolation tree differentiates each data from the others. The anomalous instances [36] are near the root node of the tree compared to the normal data [37], owing to their high sensitivity to segmentation, therefore the anomalous instances can be discovered with a limited number of distinctive circumstances.

Creating an iTree-based forest is the purpose of the iForest algorithm. The length of the isolation tree can be defined in the Isolation Forest algorithm [38] for easy presentation and estimation.

Consider I to be an isolation tree node. I is either an exterior node without a child or an interior node with exactly two daughter nodes and one test (Il, Ir). According to the value of the property and the size of the split values, the data points are divided into Il and Ir. To make an isolation tree, pick an attribute Z and a split value V at random from the data set S = s_1, s_2, s_n, and then divide every data object i.e., s_i by the value of its attribute Z (named di(Z)). If di(Z) < v, then it is kept in the left subtree and vice versa. Iterative construction of the left and right subtrees is done till any one of the following requirements is completed: a. There is just one data set or multiple data sets that are identical in S; b. when di(Z)e reaches its highest point.

Path Length: Definition

The number of edges traversed by a point p in an isolation tree from the root node until the traversal is ended at an external node is calculated by G(p). The path length of the failed query in the binary search tree given the data set S:

$$C(n) = 2G(n - 1) - (2(\mathrm{n} - 1)/\mathrm{n}), \tag{6.8}$$

where G(i) is the harmonic number, and it can be estimated by In(i) + 0.5772156649 (Euler's constant). The number of leaves is denoted by the letter n. We use c(n) to normalize g(p) because it is the average of g(d) given n. An instance p's anomaly score s is defined as:

$$S(p, n) = 2 - E(h(p))c(n), \tag{6.9}$$

where,

E(g(p))—average of g(p) from a assembly of isolation trees.

The iForest algorithm creates an iForest from a specified number of iTree. To ensure iTree diversity, subsets of S are obtained via randomization and used to construct each isolated Tree. To find the leaf node of the data s, traverse the isolation Tree collection in the isolation Forest. The abnormal proportion of s is then estimated based on the length of its route, and the abnormal evaluation of s is performed.

The iForest algorithm [39] creates an iForest from a specified number of iTree [40]. To ensure iTree diversity, subsets of S are obtained via randomization and used to construct each isolated Tree. To find the leaf node of the data s, traverse the isolation Tree collection in the isolation Forest. The abnormal proportion of s is then estimated based on the length of its route, and the abnormal evaluation of s is performed.

6.3.2 Deep Learning-Based Anomaly Detection

(1) Multilayer Perceptron (MLP)

Let us consider a thermal power plant where timestamped sensor [41] log data $\left(D = d^1, d^2, d^3, \ldots, d^T\right), T > 0$ is generated with 'n' actuators and 'm' sensors and with each discrete time step $t \in [1, 2, \ldots, T]$ the data is recorded. A two-tuple vector $d^{\,t} = <s^{\,t}, n^{\,t}>$ represents each element of the data D, where the readings of the sensors which are continuous are given by $s^{t} \in R^{m}$ and for the actuators, the categorical readings are depicted by $n^{t} \in k^{n}$.

(a) Time Series Prediction Using MLP

Another case where anomaly detection is used is for the prediction or detecting anomalies in multivariate time series data [42]. Let us consider a time series as shown in Eq. (6.10) with 'n' dimensions for every step of time $t \in [1, 2, \ldots, T]$

$$d^t = \left\{x_1^t, x_i^t, x_k^t\right\};\, xit \in R^k, \quad \forall i, 1 \le i \le k. \tag{6.10}$$

In the above equation, the number of actuators and sensors is represented by k and are also called tags and the observed data is depicted by d^t which lies in the range of i th tag to tth step of the time. A data mapping technique [43] must be used for data conversion of actuators as they are categorical. Thus, this mapping technique will convert the categorical actuator readings into numerical values [44]. Commonly, the window method is used for detecting anomalies in thermal power plants.

Consider the lengths of the input window (I_{w}) and output window (F_{w}) as $L \in N$ and $\widehat{L} \in N, L < \widehat{L}$ respectively.

$$I_{\text{w}} = \left(d^t, d^{t+1}, d^{t+2}, \ldots, d^{t+L}\right);\, L \in N \tag{6.11}$$

$$F_w = \left(\hat{d}^t, \hat{d}^{t+1}, \hat{d}^{t+2}, \ldots, \hat{d}^{t+L}\right);\, \hat{L} \in N \tag{6.12}$$

The number of input and output windows is given by $\left\lfloor (T - L)/\hat{L} \right\rfloor$ as the log of the data 'D' is noted every step of time 'T'. A function such as a non-linear transformation one let's assume 'f' relates Eqs. (6.11) and (6.12) in the following way by using a model [45] of neural network [46]

$$F_w = f(I_w),\ f : R^{L\times k} \rightarrow R^{\hat{L}xk}. \quad (6.13)$$

By analyzing Eq. (6.13), we can learn a model 'f' by using the system state's input window (I_{w}) for forecasting every sensor's behavior with the least error of forecasting 'E_{f}' shown in Eq. (6.14)

$$E_f = \frac{1}{k}\sum_{i=1}^{k}\left|\hat{x}_i^t - x_i^t\right| \quad (6.14)$$

where $\widehat{x}_i^t$ and x_i^t are the predicted and actual values of the tag and the time step at the ith and tth place respectively.

MLP contains multiple consecutive layers [47] which are connected in a sequence and each layer contains neurons that connect to the neurons of the next layer to form a feed-forward structure that is unidirectional. We will describe the architecture of MLP layer by layer. The 'input layer' which is the first layer of the network as shown in Fig. 6.2 feeds the inputs as state variables from a thermal power plant. The 'output layer' which is the final layer of the architecture predicts the behavior of the system. There exist some layers in the middle of the input and the output layer known as 'hidden layers' where each layer's input is obtained from its preceding layer.

Equation (6.15) represents a mathematical representation of each layer.

$$O_i^{(l)} = \emptyset\left(\sum_{k=1}^{n_l} O_j^{(l-1)} * W_{ki}^{(l)} * W_i^{(l)}\right);\ 1 \le l \le L \quad (6.15)$$

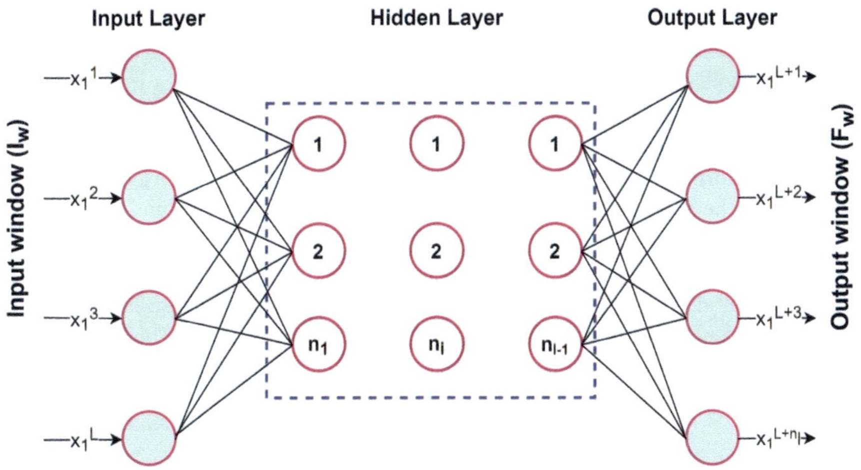

Fig. 6.2 Architecture of multi-layer perceptron

In Eq. (6.15), n_l is the number of neurons in the lth layer and L depicts the non-input layers. $W_i^{(l)}$ and $W_{ki}^{(l)}$ represent the weights of connection between the ith neuron in the lth and the $(l-1)$th layer and bias of the lth neuron in the lth layer, respectively. We denote the activation function as $\varnothing()$ which is configured as ReLU as we deal with the state variable which is non-negative. Lastly, the length n_l of the output layer i.e., $l = L$ coincides with the overall network result.

(2) PNN

Kernel discriminant analysis (KDA), which is a statistical algorithm [48] that is implemented by probabilistic neural networks (PNN) [49]. The foundation of the probabilistic networks is based upon the estimation of the probability density function (PDF) and Bayesian classification theory. Multiple subnetworks form a PNN where each network is a Parzen window pdf estimator belonging to every class. PNN's improvement or development is dependent on the development of the Parzen window classifiers. A nonparametric procedure that produces an estimation of a PDF by super positioning several windows or copies of a function is known as the Parzen windows method. By using the training data [50], a classification result is obtained by the Parzen windows classifier once the calculation of the PDF is done for every class. The result or decision of the multicategory classification is depicted below

$$p_k f_k > p_j f_j \tag{6.16}$$

In the above equation, f_k represents the estimated PDF of k class and p_k depicts the earlier probabilities when such examples from class k occurred. There are four layers as shown in Fig. 6.3 namely, summation layer, pattern layer, input layer, and decision layer in the architecture of PNN. 'c' classes are recognized by the structure of PNN and 'n' features are shown by the first layer. The total number of training instances and the total number of nodes that are in the pattern layer is the same. The number of nodes in the summation layer and the number of classes in the training instances is the same. The role of the input layer is to just distribute the input to the neurons in the pattern layer which is partially connected to the summation layer and does not perform any kind of computation. The pattern units relate to a classification from which the pattern of training was chosen, and the role of summation units is to sum these inputs.

Training vectors are given to the PNN network and the PNN builds a set of multivariate probability densities derived from these vectors. The pattern layer receives the input instance with an unknown category through propagation. The output of the node is computed after every node receives the input in the pattern layer.

$$\pi_i^c = \frac{1}{(2\pi)^{n/2\sigma^n}} e^{-\frac{(x-x_{ij})^T (x-x_{ij})}{2\sigma^2}} \tag{6.17}$$

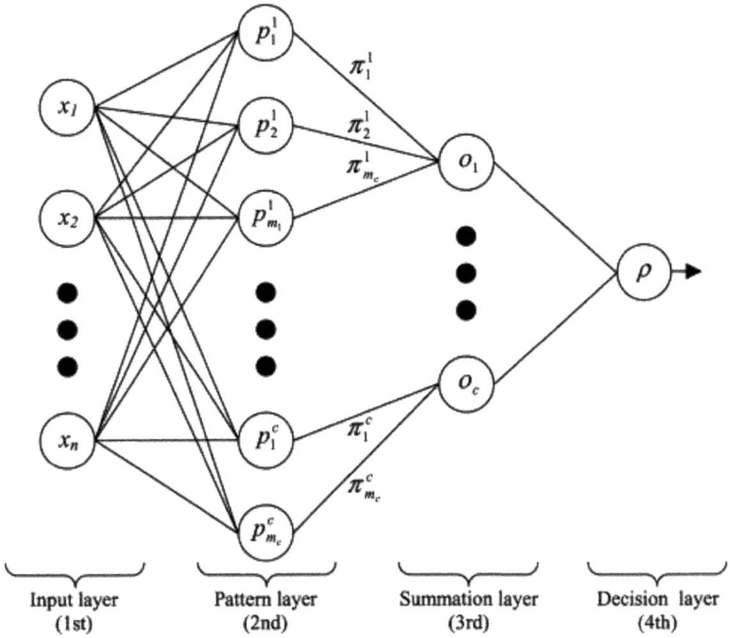

Fig. 6.3 Illustration of a probabilistic neural network [52]

In the above Eq. (6.17), σ is a smoothing parameter, 'n' represents the number of attributes of the input instance 'x', and x_{ij} is a training instance that corresponds to the 'c' category.

The neurons of the summation layer summarize and averages the output of all neurons [51] to calculate the maximum likelihood of pattern x being classified into class 'c' as shown in Eq. (6.18).

$$p_i(x) = \frac{1}{(2\pi)^{n/2}\sigma^n}\frac{1}{N_i}\sum_{i=1}^{N_i} e^{\frac{(x-x_{ij})^T(x-x_{ij})}{2\sigma^2}} \tag{6.18}$$

In the above equation, the total number of samples in class c is depicted by N_i. The decision layer unit classifies the pattern x according to the Bayes' decision rule based on the output of all the neurons of the summation layer. This occurs only if the losses related to making a wrong decision for each class are the same and the a priori probability for every class is the same.

$$C(x) = argmax\{p_i(x)\}, i = 1, 2, \ldots, c \tag{6.19}$$

In the above equation, m represents the total number of classes in the training samples and $C(x)$ represents the estimated class of the pattern x. If the losses related to making a wrong decision are different for every class and the a priori probabilities are also different for every class, then the neurons of the summation layer give the following output

$$C(x) = argmax\{p_i(x)cost_i(x)apro_i(x)\}, i = 1, 2, \ldots, c \qquad (6.20)$$

The cost for incorrectly classifying the input vector is denoted by $cost_i(x)$ and in class c, the prior probability of occurrence of patterns is denoted by $apro_i(x)$

6.4 Conclusion

In this book chapter, we explained what anomalies are and why it is essential to detect these anomalies in a thermal power plant which are equipped with different sensors. We discussed the importance of monitoring the condition of a thermal power plant and its predictive maintenance. Furthermore, we discussed how AI is being used to detect anomalies in thermal power plant systems. We identified and explained the working of various machine learning and deep learning algorithms which are being used to detect anomalies in the sensor data of these thermal power plants.

References

1. T. Bhardwaj, R. Mittal, H. Upadhyay, L. Lagos, *Applications of Swarm Intelligent and Deep Learning Algorithms for Image-Based Cancer Recognition* (2022), pp. 133–150
2. P. Hundi, R. Shahsavari, Comparative studies among machine learning models for performance estimation and health monitoring of thermal power plants. Appl. Energy **265**(August 2019), 114775 (2020). https://doi.org/10.1016/j.apenergy.2020.114775
3. M.J. Montes, A. Abánades, J.M. Martínez-Val, Performance of a direct steam generation solar thermal power plant for electricity production as a function of the solar multiple. Sol. Energy **83**(5), 679–689 (May2009). https://doi.org/10.1016/j.solener.2008.10.015
4. L. Banjanovic-Mehmedovic, A. Hajdarevic, M. Kantardzic, F. Mehmedovic, I. Dzananovic, Neural network-based data-driven modelling of anomaly detection in thermal power plant. Automatika **58**(1), 69–79 (2017). https://doi.org/10.1080/00051144.2017.1343328
5. T. Bhardwaj, S.C. Sharma, *Internet of Things: Route Search Optimization Applying Ant Colony Algorithm and Theory of Computation* (2015), pp. 293–304
6. J.F. Olesen, H.R. Shaker, Predictive maintenance for pump systems and thermal power plants: state-of-the-art review, trends and challenges. Sensors (Switzerland) **20**(8) (2020). https://doi.org/10.3390/s20082425
7. Y. Zhang, Z.Y. Dong, W. Kong, K. Meng, A composite anomaly detection system for data-driven power plant condition monitoring. IEEE Trans. Ind. Inform. **16**(7), 4390–4402 (2020). https://doi.org/10.1109/TII.2019.2945366
8. T. Bhardwaj, S.C. Sharma, Cloud-WBAN: an experimental framework for cloud-enabled wireless body area network with efficient virtual resource utilization. Sustain. Comput. Inform. Syst. **20**, 14–33 (2018). https://doi.org/10.1016/j.suscom.2018.08.008
9. P. Gangwani, A. Perez-Pons, T. Bhardwaj, H. Upadhyay, S. Joshi, L. Lagos, Securing environmental IoT data using masked authentication messaging protocol in a DAG-based blockchain: IOTA tangle. Futur. Internet **13**(12), 312 (2021). https://doi.org/10.3390/fi13120312

10. J. Chen, H. Li, D. Sheng, W. Li, A hybrid data-driven modeling method on sensor condition monitoring and fault diagnosis for power plants. Int. J. Electr. Power Energy Syst. **71**, 274–284 (2015). https://doi.org/10.1016/j.ijepes.2015.03.012
11. J. Soni, N. Prabakar, J. Kim, Prediction of component failures of telepresence robot with temporal data, in *30th Florida Conference on Recent Advances in Robotics May 11–12, 2017, Florida Atlantic University, Boca Raton, Florida*, no. 1 (2017), pp. 2–5
12. S.K. Yang, A condition-based preventive maintenance arrangement for thermal power plants. Electr. Power Syst. Res. **72**(1), 49–62 (2004). https://doi.org/10.1016/j.epsr.2004.03.007
13. Y. Sheiretov, D. Grundy, V. Zilberstein, N. Goldfine, S. Maley, MWM-array sensors for in situ monitoring of high-temperature components in power plants. IEEE Sens. J. **9**(11), 1527–1536 (2009). https://doi.org/10.1109/JSEN.2009.2019335
14. P. Gangwani, J. Soni, H. Upadhyay, S. Joshi, A deep learning approach for modeling of geothermal energy prediction. Int. J. Comput. Sci. Inf. Secur. **18**(1), 62–65 (2020)
15. D. Gangwani, Q. Liang, S. Wang, X. Zhu, An empirical study of deep learning frameworks for melanoma cancer detection using transfer learning and data augmentation, in *2021 IEEE International Conference on Big Knowledge (ICBK)* (2021), pp. 38–45. https://doi.org/10.1109/ICKG52313.2021.00015
16. J. Soni, N. Prabakar, H. Upadhyay, Behavioral analysis of system call sequences using LSTM Seq-Seq, Cosine similarity and Jaccard similarity for real-time anomaly detection, in *2019 International Conference on Computational Science and Computational Intelligence (CSCI)* (2019), pp. 214–219. https://doi.org/10.1109/CSCI49370.2019.00043
17. T. Bhardwaj, H. Upadhyay, L. Lagos, *Deep Learning-Based Cyber Security Solutions for Smart-City: Application and Review* (2022), pp. 175–192
18. P. Rathore, J. Soni, N. Prabakar, M. Palaniswami, P. Santi, Identifying groups of fake reviewers using a semisupervised approach. IEEE Trans. Comput. Soc. Syst. **8**(6), 1369–1378 (2021). https://doi.org/10.1109/TCSS.2021.3085406
19. J. Soni, N. Prabakar, H. Upadhyay, Visualizing high-dimensional data using t-distributed stochastic neighbor embedding algorithm, in *Principles of Data Science* (2020), pp. 189–206
20. D. Gangwani, P. Gangwani, Applications of machine learning and artificial intelligence in intelligent transportation system: a review, in *Lecture Notes in Electrical Engineering* (Springer, 2021), pp. 203–216
21. J. Soni, N. Prabakar, Effective machine learning approach to detect groups of fake reviewers, in *Proceedings of the 14th International Conference on Data Science (ICDATA'18), Las Vegas, NV* (2018), pp. 74–78
22. S. Joshi, H. Upadhyay, L. Lagos, N.S. Akkipeddi, V. Guerra, Machine learning approach for malware detection using random forest classifier on process list data structure, in *Proceedings of the 2nd International Conference on Information System and Data Mining—ICISDM '18* (2018), pp. 98–102. https://doi.org/10.1145/3206098.3206113
23. T. Bhardwaj, S.C. Sharma, An autonomic resource provisioning framework for efficient data collection in cloudlet-enabled wireless body area networks: a fuzzy-based proactive approach. Soft Comput. **23**(20), 10361–10383 (2019). https://doi.org/10.1007/s00500-018-3587-x
24. H. Upadhyay, L. Lagos, S. Joshi, A. Szilagyi, *Artificial Intelligence Application to D and D—20492* (2020) [Online], https://www.osti.gov/biblio/23030565
25. J. Soni, S.K. Peddoju, N. Prabakar, H. Upadhyay, Comparative analysis of LSTM, one-class SVM, and PCA to monitor real-time malware threats using system call sequences and virtual machine introspection, in *Lecture Notes in Electrical Engineering* (2021), pp. 113–127
26. C. Schuldt, L. Barbara, S. Stockholm, Recognizing human actions: a local SVM approach ∗ Dept. of Numerical Analysis and Computer Science, in *Proceedings of the 17th International Conference on Pattern Recognition, ICPR 2004,* vol. 3 (2004), pp. 32–36
27. B. Schölkopf, R. Williamson, A. Smola, J. Shawe-Taylor, J. Piatt, Support vector method for novelty detection. Adv. Neural Inf. Process. Syst. 582–588 (2000)
28. K. Bhanot, S.K. Peddoju, T. Bhardwaj, A model to find optimal percentage of training and testing data for efficient ECG analysis using neural network. Int. J. Syst. Assur. Eng. Manag. **9**(1), 12–17 (2018). https://doi.org/10.1007/s13198-015-0398-7

29. T. Bhardwaj, H. Upadhyay, S.C. Sharma, *An Autonomic Resource Allocation Framework for Service-Based Cloud Applications: A Proactive Approach* (2020), pp. 1045–1058
30. V. Hodge, J. Austin, A survey of outlier detection methodologies. Artif. Intell. Rev. **22**(2), 85–126 (2004). https://doi.org/10.1023/B:AIRE.0000045502.10941.a9
31. K.L. Li, H.K. Huang, S.F. Tian, W. Xu, Improving one-class SVM for anomaly detection. Int. Conf. Mach. Learn. Cybern. **5**(November), 3077–3081 (2003). https://doi.org/10.1109/icmlc.2003.1260106
32. J. Soni, N. Prabakar, H. Upadhyay, *Feature Extraction Through Deepwalk on Weighted Graph* (2019)
33. F.T. Liu, K.M. Ting, Z.H. Zhou, Isolation forest, in *Proceedings of the IEEE International Conference on Data Mining, ICDM* (2008), pp. 413–422. https://doi.org/10.1109/ICDM.2008.17
34. H. Upadhyay, L. Lagos, S. Joshi, A. Abrahao, *Big Data Framework with Machine Learning for D and D Applications—19108* [Online] (2019), https://www.osti.gov/biblio/23002927
35. S. Joshi, H. Upadhyay, L. Lagos, *Deactivation and Decommissioning Web Log Analysis Using Big Data Technology—15710* [Online] (2015), https://www.osti.gov/biblio/22824525
36. D. Xu, Y. Wang, Y. Meng, Z. Zhang, An improved data anomaly detection method based on isolation forest, in *Proceedings of the 2017 10th International Symposium on Computational Intelligence and Design (ISCID)*, vol. 2 (2018), pp. 287–291. https://doi.org/10.1109/ISCID.2017.202
37. T. Bhardwaj, H. Upadhyay, S.C. Sharma, Autonomic resource allocation mechanism for service-based cloud applications, in *2019 International Conference on Computing, Communication, and Intelligent Systems (ICCCIS)* (2019), pp. 183–187. https://doi.org/10.1109/ICCCIS48478.2019.8974515
38. T. Bhardwaj, S.C. Sharma, Fuzzy logic-based elasticity controller for autonomic resource provisioning in parallel scientific applications: a cloud computing perspective. Comput. Electr. Eng. **70**, 1049–1073 (2018). https://doi.org/10.1016/j.compeleceng.2018.02.050
39. T. Bhardwaj, H. Upadhyay, S.C. Sharma, Service-based cloud applications: a queuing—model based approach, in *2020 10th International Conference on Cloud Computing, Data Science & Engineering* (2020), pp. 605–610
40. S. Hariri, M.C. Kind, R.J. Brunner, Extended isolation forest. IEEE Trans. Knowl. Data Eng. **33**(4), 1479–1489 (2021). https://doi.org/10.1109/TKDE.2019.2947676
41. T. Bhardwaj, S. Chander Sharma, An efficient elasticity mechanism for server-based pervasive healthcare applications in cloud environment, in *2017 IEEE 19th International Conference on High Performance Computing and Communications Workshops (HPCCWS)* (2017), pp. 66–69. https://doi.org/10.1109/HPCCWS.2017.00016
42. P. Esling, C. Agon, Time-series data mining. ACM Comput. Surv. **45**(1), 1–34 (2012). https://doi.org/10.1145/2379776.2379788
43. M. Atay, A. Chebotko, D. Liu, S. Lu, F. Fotouhi, Efficient schema-based XML-to-relational data mapping. Inf. Syst. **32**(3), 458–476 (2007). https://doi.org/10.1016/j.is.2005.12.008
44. T. Bhardwaj, H. Upadhyay, S.C. Sharma, Framework for quality ranking of components in cloud computing: regressive rank, in *Proceedings of Confluence 2020—10th International Conference on Cloud Computing, Data Science & Engineering* (2020), pp. 598–604. https://doi.org/10.1109/Confluence47617.2020.9058016
45. S.S. Sirigineedi, J. Soni, H. Upadhyay, Learning-based models to detect runtime phishing activities using URLs, in *Proceedings of the 2020 the 4th International Conference on Compute and Data Analysis* (2020), pp. 102–106. https://doi.org/10.1145/3388142.3388170
46. N.V. Dharwadkar, A.A. Dixit, A.K. Kannur, M.A.B. Kadampur, S. Joshi, Identification of reasons behind infant crying using acoustic signal processing and deep neural network for neonatal intensive care unit. Int. J. Inf. Retr. Res. **12**(1), 1–17 (2022). https://doi.org/10.4018/IJIRR.289576
47. G. Raman, N. Somu, A.P. Mathur, A multilayer perceptron model for anomaly detection in water treatment plants. Int. J. Crit. Infrastruct. Prot. **31**, 100393 (2020). https://doi.org/10.1016/j.ijcip.2020.100393

48. T. Bhardwaj, M. Kumar, S.C. Sharma, *Megh: A Private Cloud Provisioning Various IaaS and SaaS* (2018), pp. 485–494
49. S.G. Wu, F.S. Bao, E.Y. Xu, Y.X. Wang, Y.F. Chang, Q.L. Xiang, A leaf recognition algorithm for plant classification using probabilistic neural network, in *ISSPIT 2007—2007 IEEE International Symposium on Signal Processing and Information Technology* (2007), pp. 11–16. https://doi.org/10.1109/ISSPIT.2007.4458016
50. H. Upadhyay, L. Lagos, S. Joshi, E. Mohammed, K. Cooper, Predictive analytics with big data-spark framework. Nucl. Plant J. **30**(2) (2012) [Online], https://digitaleditions.nuclearplantjournal.com/MA18/29/
51. B.M. Ozyildirim, M. Avci, Generalized classifier neural network. Neural Netw. **39**, 18–26 (2013). https://doi.org/10.1016/j.neunet.2012.12.001
52. A. Hajdarevic, I. Dzananovic, L. Banjanovic-Mehmedovic, F. Mehmedovic, Anomaly detection in thermal power plant using probabilistic neural network, in *2015 38th International Convention on Information and Communication Technology, Electronics and Microelectronics MIPRO 2015—Proceedings*, no. May (2015), pp. 1118–1123. https://doi.org/10.1109/MIPRO.2015.7160443

Chapter 7
Deep Learning Aided Intrusion Detection for Smart Grid Cyber Security

MD. Rifat Hossain and Tushar Bhardwaj

Abstract Smart grid is a convoluted cyberphysical system and the evident future of modern power grid. Conventional power structure accumulates unidirectional power flow with minimal control. The arithmetic of smart grid allows bi-directional flow of information with smart sensory devices and cloud-based services. The connected structure, while results in increased control, quality, and reliability aspects, allows cyber attackers to impair the system dynamics. Artificially intelligent models can process large amounts of data, learn from previous instances, and formulate strategies to predict system states in real-time with minimal human intervention. For intrusion detection in the data traffic, this paper analyzes Deep Learning (DL) models, formed with three hidden layers and three different optimizer functions. The model performance is analyzed by forming the Confusion Matrix and calculating precision, recall, accuracy, and F-1 score. It has been found that, in all the instances, the DL model performs satisfactorily to predict actual system state based on data traffic. Among all the iterations, the highest levels of precision, recall, accuracy, and F-1 score achieved are 98.23, 85.68, 97.55, and 92.91% respectively. Consequently, neural networks are able to promptly detect system dynamics with extensive training, a well-formed network architecture, and adequate system information.

Keywords Smart grid · Cyberphysical system · Deep learning · Intrusion detection system · Performance indices

---

MD. R. Hossain (✉)
Electrical and Computer Engineering, Florida International University, Miami, FL, USA
e-mail: mhoss058@fiu.edu

T. Bhardwaj
Applied Research Center, Florida International University, Miami, FL, USA
e-mail: tbhardwa@fiu.edu

T. Bhardwaj et al. (eds.), *Artificial Intelligence in Cyber Security: Theories and Applications*, Intelligent Systems Reference Library 240,
https://doi.org/10.1007/978-3-031-28581-3_10

7.1 Introduction

Modern civilization is undergoing extensive transformation due to the swelling inclusion of cyber-physical systems (CPS) in our day-to-day lives. The emerging smart grid (SG) technology is one of the most involuted CPS of the present era. With the advent of the digital era the need for electrical energy is increasing abruptly. Today's energy needs cannot be successfully met by the conventional electricity grid. On the one hand, fossil fuel resources are depleting, resulting in rapid integration of renewable power sources. On the other hand, pioneering technologies such as automated manufacturing, electric vehicles, and microprocessor based devices are continuously replacing the traditional processes with a need of precise power quality [1, 2]. All aspects of the power system including generation, transmission, and distribution are going through a sharp modification. The capacity to measure, communicate, and control power systems has stretched to unprecedented limits. Precise information is being collected about the system parameters for time intervals of less than a second to strengthen decision-making in pseudo-real time [3]. The term Smart Grid (SG) refers to the joint avenue of the upgraded power system structure with Advanced Metering Infrastructure for meeting operational precision [4–6].

For conventional power grids, only energy flows from the generation side to the load side. The metering is done at the consumer premises. Alternatively, in the smart grid structure two-way flow is established. Synonymous to a conventional grid, the power flows to the consumer end where smart meters are installed. The smart meters send information in the reverse direction to a central station for establishing a real-time control/monitoring network from the generation side. The Supervisory Control and Data Acquisition (SCADA) technology has been integrated into the SG as a mean to ensure security of bi-directional communication links. Additionally, several bi-directional communication protocols are adopted by the American National Standard Institute for safeguarding the process [7–9]. The immergence of the SG technology has been marked in the adoption of 100 SG projects in the American Reinvestment and Recovery Act with a total cost of USD 3.4 Billions [10]. The National Institute of Standards and Technology (NIST) concedes the SG model which is depicted in Fig. 7.1. The figure highlights operation of control centers to ensure the secure and optimal flow of energy to the load.

From a holistic view, SG, as a single concept, amalgamates the dynamism and complexity of standalone energy, control, sensory, supervisory, and financial systems. The inherent diversity of the SG's architecture has raised alarming vulnerabilities and generic apprehension about security and resilience of the real-life system along with the metering infrastructure [12–14]. The SG security issues can be classified into two main branches- physical and cyber security [15]. An intrusion that affects either the physical power system or the communication network or both should be detected as the smart grid network is a hybrid of the two. These issues combinedly comprise the cyber-physical domain of smart grid [16]. Although several physical attacks on the SG are recorded over time, the cyber-attacks are more alarming, sophisticated, and inflating compared to the physical attacks [17–20]. Whereas advanced technologies

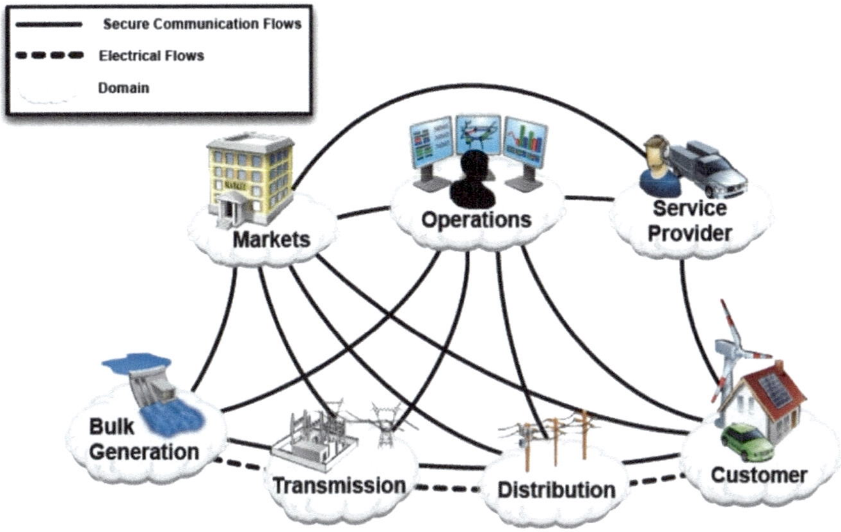

Fig. 7.1 Conceptual model of smart grid [11]

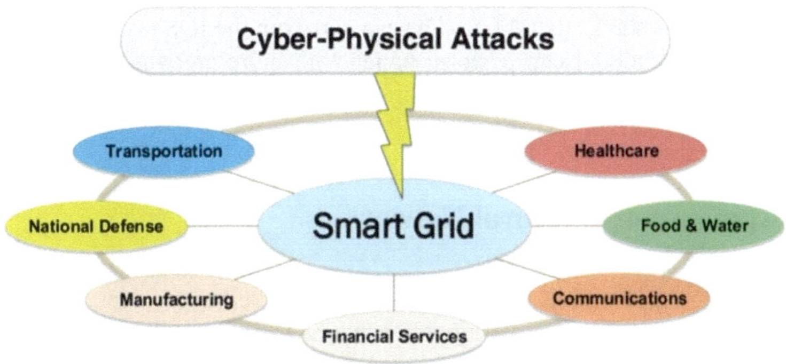

Fig. 7.2 Cyber-physical attack impacts on smart grid [22]

for sensing, computation, and data transfer have made many modern applications of the SG a reality, as a trade-off, power grids today suffer from cyber-threats due to their substantial dependence on cloud-based services [21].

It is of paramount importance to detect and rectify cyber intrusion into the SG architecture due to the interdependency of individual sectors to the power grid for fulfilling their power needs. The essential connectivity to the SG poses threat of malfunctioning for all the major sectors if a cyber-attack is carried out on the SG. The situation is represented using Fig. 7.2.

It can be observed that all the major national infrastructures can be impacted through SG cyber-attacks. Furthermore, the future goal of smart grid is to make it more reliable, economical, and efficient in an environment of increasing power demand while incorporating a tremendous increase in data communications [23, 24]. It is therefore of critical importance to ensure cyber security of the Smart Grid [25].

For real-time detection and rectification of SG cyber attacks, artificial intelligence (AI) aided intrusion detection systems (IDS) can be utilized [26, 27]. Neural network (NN) is an AI model and corresponds to the deep learning (DL) domain, which is a subset of machine learning (ML). DL differentiate from ML in the layered structure used during its' formation [28]. NN can be defined as a computationally aware network assembled with an array of simple, highly interconnected neurons [29]. The idea behind neural networks is to mimic the structure of the human brain based on the fact that humans can learn by seeing many examples [30]. Similarly, while creating a NN model, no specific instructions are given to the program for execution of a work. Rather, numerous examples are given to the model during training to tune model parameters for matching with the desired outcome. In this study, neural network models are used to perform IDS tasks for the cyber-physical domain security of SG. The architecture of the NN model is formed using convolutional type layers. Thus, the IDS model can be denoted as a convolutional neural network (CNN) as well. The analysis shows that AI models can carry out the IDS task with reasonable performance. The model can perform better with more amount of training, more integrated model architecture, and fine-tuned model parameters [31, 32].

7.2 Application of Neural Network

Originally, NN applications root in the computer science domain, but almost all branches of modern studies have adopted machine learning or deep learning models for precise decision making. Pattern recognition, optimal decision-making, speech recognition, time series forecast, predictive and automatic control are among the heavily deployed NN fields [33, 34]. In the recent era, researchers are extensively implementing NN topologies in various power system aspects. Satisfactory results can be seen for NN models developed in the field of power system operation and control [35, 36], load forecast [37–39], cost effectiveness [40], fault and security analysis [41–44], transient and voltage stability [45–47], etc.

The analysis of real-time sensory data has been carried out using ML, fuzzy logic, data mining technique, NN, support vector machine, and other numerical analysis [48–50]. The smart grid stability analysis has been a specialized implementation case of DL models [51]. During the formation of a DL model, many different hyperparameters are needed to be selected for proper fitting of the model layout with the problem dynamics [52]. Also, as for the DL model, the input–output relationships are not specifically provided, the model performance depends on the training dataset

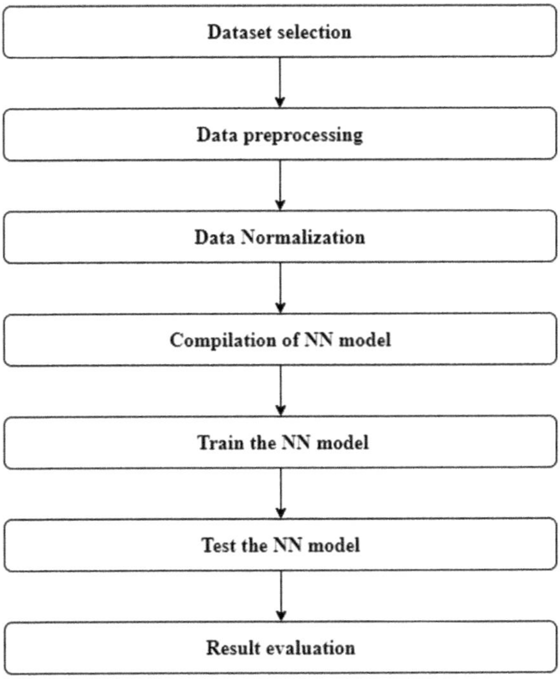

Fig. 7.3 Neural network model application steps

quality [53]. The process involved in founding the IDS using CNN model is shown in Fig. 7.3.

A. *Dataset Selection*

Several datasets are being used by researchers for the purpose of developing machine learning or deep learning models to detect CPS intrusions. Studies have been conducted that have analyzed synthetic datasets as well as those that have analyzed actual field data [54]. Usually, however, the data used is not publicly available [55]. For intrusion detection of CPS the following publicly available datasets have been widely used for research purposes [56]:

- DARPA Intrusion Detection datasets [57–59] were published by MIT Lincon Laboratories. The datasets include manually created intrusions on network traffic. Various intrusions are created and corresponding changes in system dynamics are noted during a five-week study for creating the dataset.
- The KDDCUP'99 dataset [60] was formed based on the listings of the DARPA datasets. This dataset was first published in 1999 at Knowledge Discovery and Data Mining (KDD) workshops at the Association for the Advancement of Artificial Intelligence (AAAI) conferences. Although the KDDCUP'99 dataset has issues such as record redundancy, it has been the most widely used dataset for intrusion detection on network traffic research.

- NSL-KDD dataset [61] is an upgraded and more realistic version of KDDCUP'99. It was published in 2009. In the NSL-KDD dataset the record redundancy issue of KDDCUP'99 has been rectified. Moreover, a more precise selection has been made to prepare a dataset with uniformly distributed system states [62].

NSL-KDD dataset has been used for the analysis in this paper. In total, 125,973 entries are included in the full NSL-KDD training dataset. Among the entries, 67,343 data points correspond to the normal class, which means no malicious activity is present in the network traffic. The total number of attack entries in the dataset is 58,630, of which DoS comprises the major portion with a total 45,927 entries. The Probe, U2R, and R2L attack types respectively have 11,656, 52, and 995 records [62].

Each entry in the training and testing datasets has 41 feature values and one class attribute. The class attribute reveals the status of the system, which can be either in normal state or in under-attack state. Furthermore, whereas the training set contains 22 different attack types, the testing dataset contains 17 unique attacks in addition to the 22 previously mentioned attacks in the training dataset. All 39 attack types fall under four major attack categories. The major categories are Denial of Service (DOS), Probe, Remote to Local (R2L), and User to Root (R2L).

The main four attack categories are described in Table 7.1 along with the attack types which fall under each category and number of attacks in the training and testing datasets.

B. *Data Preprocessing*

Usually, data preprocessing is carried out prior to implementing a deep learning algorithm. A validation procedure is required because the dataset might contain duplicate data, missing values, or erroneous entries [63, 64]. For the machine to properly read all the different feature entries, they need to be in a generic format as well. The selected NSL-KDD dataset contains both string and numeric type data. The features pertaining to string type value are 'protocol_type', 'service', and 'flag'. Other features are either integer or float type data. The class attribute value of the dataset is also string type. The class attribute primarily divided into five classes, which are normal, DoS, probe, R2L, and U2R. The last four classes represent distinct attack sets, which are merged during the simulation to be categorized as under-attack system state.

Furthermore, the NSL-KDD dataset only contains the values of features and classes. The feature and class names excluded for diverse use of the dataset. In the preprocessing stage, the names of the features and class are added separately to the dataset.

C. *Data Normalization*

Data normalization is an essential prerequisite for acquiring unbiased results from a dataset. This step is important because different attributes in the training and testing dataset have different value ranges [65]. If all attribute values are not normalized to

Table 7.1 Attack categories of NSL-KDD dataset

Attack category	Description	Attack types	Train set records	Test set records
DoS	By disrupting the services of a host connected to a network, malicious actors attempt to make a machine or network resource unavailable to its intended users	Apache2, mailbomb, processtable, udpstorm, back, land, Neptune, pod, smurf, teardrop	45,927	7458
Probe	Network scanning has been done to acquire system information and to identify potential weaknesses, mostly with the intention of launching a future cyberattack	mscan, saint, Ipsweep, nmap, portsweep, satan	11,656	2421
R2L	Illegitimate data package forwarding over a network by gaining local user access without a system user account in that network	httptunnel, named, sendmail, snmpgetattack, xlock, xsnoop Spy, warezclient, ftp_ write, guesspasswd, imap, multihop, phf, warezmaster	995	2754
U2R	Network misuse by an assailant with regular user profile to acquire root user privileges	Bufferoverflow, loadmodule, perl, rootkit, ps, snmpguess, sqlattack, worm, xterm	52	200

a uniform margin, attributes with an upper value range will receive a higher significance in decision making than attributes with lower values. As all attributes in a cyberphysical network are individually important for ensuring security, data normalization has been implemented to avoid biased decision-making by the neural network model. Using Z-score normalization of attribute values, this study aims to achieve the discussed goal. Z-score normalization, also termed zero mean normalization, processes the attribute values in a way that their mean becomes zero and standard deviation becomes one.

D. *Compilation of NN Model*

NN can be defined as a collective statistical learning tool which analyzes large amounts of training inputs to provide estimated functional outputs. The basic building block of NN is called a perceptron or a neuron. Each layer of a NN is a combination of different numbers of perceptrons. The generic NN architecture comprises an input layer, single or multiple hidden layers, and an output layer. In each layer, the forming perceptrons receive inputs from the previous layer's perceptrons. The aggregated inputs are multiplied by an assigned weight, and then passed into an activation function to produce an output. An array of activation functions for NN layers can be found including the logistic function, a trigonometric function, a step function etc. The activation function determines the behavior of the model to correct itself during the training process. After each iteration during training, the NN model adjusts the associate weights as a corrective measure starting from the output layer to the input layer. This process is called back propagation. The aim of back propagation is to be more aware of the data dynamics and make better predictions. The loss and optimizer functions play an essential role in selecting weight adjustments throughout the network.

In our study, the platform Jupyter Notebook, v 6.3.0, and language Python 3 have been used for implementation of the NN model. Once the data has been preprocessed and normalized different NN models have been formed for fetching the training dataset and analyzing their performance on the test dataset. For the NN models, the loss function is selected to be 'Categorical_Crossentropy'. Also, both 'Maxpooling'/ 'averagepooling' and dropout functions have been deployed on hidden convolution layers to reduce computational times. The 'softmax, activation function is chosen for the output layer and the relu activation function is adopted for the input and hidden layers. The varying features of the NN models are formation of hidden layers, optimizer function, dropout rate, and number of iterations. Also, three different activation functions are used to inspect their impact on the decision accuracy.

E. *Training of Neural Network*

During training of the deep learning model, a reduced NSL-KDD training dataset has been used. The name of this publicly available training dataset is 'KDDTrain+_20Percent'. It contains 20% of the full training dataset and was provided by the original publishers of the NSL_KDD dataset. This training set accommodates 25,192 entries with 53.39% of them being in the normal category. The number of network intrusion entries is 11743 containing all four types of major attacks.

F. *Testing Neural Network*

The NSL-KDD testing dataset is named 'KDDTest+' and it contains 22,544 records. From the original publishers of the NSL-KDD dataset, a reduced testing dataset is also made available which is named 'KDDTest-21' and it contains 11,850 listings. In our study, the 'KDDTest+' dataset has been used for testing the generated NN model because of the larger volume of the test set. This dataset contains 9711 normal entries. Among the attack data-points DoS, Probe, U2R, and R2L have 7548, 2421, 200, and 2754 test values respectively.

7.3 Result Evaluation

Different parameters can be analyzed for evaluating the performance of deep learning models. The performance indices considered in this study precision, recall, accuracy, and F-1 score, which are among the most recognized performance indices. Finally, based on the indices values the performance of the neural network model as an IDS is justified.

Table 7.2 Formation of confusion matrix

Confusion Matrix		Prediction	
		Attack	Normal
Actual	Attack	TP	FN
	Normal	FP	TN

To quantize the mentioned parameters of the model, the confusion matrix (CM) has been formed in the python environment. It is done by importing the 'confusion_matrix' package from 'sklearn.metrics' function. The CM gives four types of possible predictions done by the NN model on the test dataset. The different prediction possibilities are True Positive (TP), False Negative (FN), False Positive (FP) and True Negative (TN). The formation of the CM is depicted in the Table 7.2.

The performance indices mentioned above are functions of the CM values and can be easily calculated using straightforward formulas. In any case, a good performance can be attributed to an AI model if the most prediction distribution falls in the TP and TN categories. Brief descriptions of the performance tokens and their corresponding calculation formulas are given in Table 7.3.

The values of the evaluation metrics acquired for different deep learning model setups are summarized in Table 7.2. The first column of Table 7.2 depicts the model architecture and represents the sequential layers which form the NN model. The first and last entry types are for the input and output layers consecutively, whereas the in-between entries render the types of the hidden layers. Two types of hidden layers are used, 'Convolution1D' (Conv1D) and 'Dense' types. The bracketed values following the layer types denote number of filters for Conv1D layers and number of neurons for Dense type layers. Additionally, for the convolution layers 'Maxpool' or 'Averagepool' (AvgPool), along with 'Dropout' functions are utilized for reducing computational burden of the model. The bracketed values adjacent to these layers represent the 'pool_size' for pool-type layers and 'Rate' for the 'Dropout' layers. To analyze model performance for different learning algorithms, three different optimizer function-based models have been formed and performance indices are calculated. The optimizer functions selected are 'Adam', 'Adadelta', and 'SGD'. Finally,

Table 7.3 Performance indices of the deep learning model

Performance indices	Description	Formula
Precision	The proportion of correctly declared positive records among all the predicted positive records in the test dataset	$\frac{TP}{TP+FP}$
Recall	The proportion of correctly predicted attack scenarios among all the actual attack scenarios in the test dataset	$\frac{TP}{TP+FN}$
Accuracy	Ratio of the correct prediction to the total number of predictions done by the model	$\frac{TP+TN}{TP+TN+FP+FN}$
F-1 score	It is the harmonic mean of precision and recall. It identifies the balance between the two parameters	$2 \times \frac{\text{Precision} \times \text{Recall}}{\text{Precision} + \text{Recall}}$

the model has been trained using increasing number of iterations, also known as epochs. The findings from model simulation are listed in Table 7.3.

Model architecture	Optimizer	Epochs	Precision	Recall	Accuracy	F-1 score
Conv1D–Conv1D (256)–MaxPool (2)–Dropout (0.2)–Conv1D (256)–MaxPool (2)–Dropout (0.2)–Conv1D (256)–AvgPool (2)–Dropout (0.2)–Dense (512)–Dense	Adam	10	89.93	67.64	77.29	77.21
		30	90.16	72.42	79.99	80.33
		50	94.30	85.68	97.55	92.91
	Adadelta	10	89.39	63.72	75.06	74.40
		30	89.74	69.00	78.41	76.82
		50	91.32	72.89	81.13	80.31
	SGD	10	97.68	57.50	75.05	72.39
		30	97.94	61.37	77.36	76.48
		50	98.23	68.02	81.51	79.98

Inspection of simulation results shows that with a greater number of epochs, the prediction results have increased steadily. The reason behind this is that with more epochs, the model gains more familiarity with the training dataset and can more precisely tune the network weights for making a better prediction on the test dataset. It can also be considered that the model performance falls relatively short of achieving 100% accuracy. It is because the number of training entries is small compared to the testing dataset. By using the full 'NSL-KDD' training dataset, the CNN outcome can be elevated. Besides, the simulation uses both pool-type and dropout functions during simulation, which curtails the model parameters by some margin, ultimately reducing the model performance. The model achievements can be further augmented by adding hidden layers and/or increasing the filter numbers/ neuron numbers of existing layers. The recall value is significantly lower in all the simulations. It highlights that the CNN models in this study, predicts positive values more accurately than negative values. Despite these limitations, the highest accuracy, precision, and F-1 score achieved during the simulation are above 90%, which suggests that the CNN is capable enough to detect attack situations in the large network traffic of cyber-physical systems.

7.4 Conclusion

The continuously evolving power system is faced with unprecedented changes, which have arisen unforeseen challenges, particularly in the cyber domain. Real-time detection requires highly responsive solutions to counteract these situations. Smart grid technology utilizes big data transfer. The vast amount of data carries information of numerous system parameters. For successful intrusion detection in the network, all data needs to be continuously monitored and real-time actions need to be carried out in case of an anomaly. In this regard, this paper flagships the effectiveness of neural network based artificial intelligence for being utilized in the scenario. The computational capability and intrinsic methodology of NNs are proven as an implementable tool during the simulation. The humongous task can be carried out by deployment of appropriately formed deep learning models. During expansion or modification of the physical system, the AI models can be easily deformed to encompass the system changes.

References

1. S.A. Aleem, S.M.S. Hussain, T.S. Ustun, A review of strategies to increase PV penetration level in smart grids. Energies **13**, 636 (2020)
2. M. Farmanbar, K. Parham, Ø. Arild, C. Rong, A widespread review of smart grids towards smart cities. Energies **12**, 4484 (2019). https://doi.org/10.3390/en12234484
3. T.S. Ustun, S.M. Farooq, S.M.S. Hussain, Implementing secure routable GOOSE and SV messages based on IEC 61850-90-5. IEEE Access **8**, 26162–26171 (2020)
4. NIST Framework and Roadmap for Smart Grid Interoperability Standards, Release 3.0, available [online]: https://doi.org/10.6028/NIST.SP.1108r3
5. 2013 IEEE PES Smart Grid Compendium, [online] available: https://www.ieee-pes.org/smart-grid-compendium-2013
6. The Smart Grid: An Introduction US Department of Energy, 2016, [online] Available: http://energy.gov/oe/downloads/smart-grid-introduction-0
7. ANSI C12.18-2006, [online] Available: https://webstore.ansi.org/StandardsINEMAIANSIC12182006R2016
8. ANSI C12.21-2006, [online] Available: https://webstore.ansi.org/StandardsINEMAIANSIC12212006R2016
9. ANSI C12.22-2008, [online] Available: https://webstore.ansi.org/StandardsINEMAIANSIC12222008
10. T.M. Chen, Smart grids, smart cities need better networks [Editor's Note], in *IEEE Network*, vol. 24, no. 2, pp. 2–3, March-April 2010. https://doi.org/10.1109/MNET.2010.5430136
11. National Institute of Standards and Technology. NIST framework and roadmap for smart grid interoperability standards, release 1.0. http://www.nist.gov/publicaffairs/releases/upload/smartgridinteroperabilityfinal.pdf.January2010
12. K. Bhat, V. Sundarraj, S. Sinha, A. Kaul, IEEE cyber security for the smart grid (2013) pp. 1–122. https://doi.org/10.1109/IEEESTD.2013.6613505
13. R. Marah, I.E. Gabassi, S. Larioui, H. Yatimi, Security of smart grid management of smart meter protection, in *2020 1st International Conference on Innovative Research in Applied Science, Engineering and Technology (IRASET)* (2020), pp. 1–5. https://doi.org/10.1109/IRASET48871.2020.9092048
14. Y. Gui, A.S. Siddiqui, S.M. Tamore, F. Saqib, Security Vulnerabilities of Smart Meters in Smart Grid, in *IECON 2019—45th Annual Conference of the IEEE Industrial Electronics Society* (2019), pp. 3018–3023. https://doi.org/10.1109/IECON.2019.8926992
15. S.M. Amin, A.M. Giacomoni, Smart grid, safe grid, in *IEEE Power and Energy Magazine*, vol. 10, no. 1 (2012), pp. 33–40. https://doi.org/10.1109/MPE.2011.943112
16. Y. Zhang, L. Wang, W. Sun, R.C. Green II., M. Alam, Distributed intrusion detection system in a multi-layer network architecture of smart grids. IEEE Trans. Smart Grid **2**(4), 796–808 (2011). https://doi.org/10.1109/TSG.2011.2159818
17. Critical Infrastructure Protection: More Comprehensive Planning Would Enhance the Cybersecurity of Public Safety Entities' Emerging Technology, Jan. 2014, [online] Available: http://www.gao.gov/assets/670/660404.pdf

18. ICS-CERT Monitor Newsletters, Aug. 2016, [online] Available: https://ics-cert.us-cert.gov/monitors
19. A. Ashok, M. Govindarasu, J. Wang, Cyber-physical attack-resilient wide-area monitoring, protection, and control for the power grid. Proc. IEEE **105**(7), 1389–1407 (2017). https://doi.org/10.1109/JPROC.2017.2686394
20. T. Bhardwaj, End-to-end data security for multi-tenant cloud environment, pp. 13–20
21. E.M. Ferragut, J. Laska, M.M. Olama, O. Ozmen, Real-time cyber-physical false data attack detection in smart grids using neural networks, in *2017 International Conference on Computational Science and Computational Intelligence* (2017)
22. H. He, J. Yan, Cyber-physical attacks and defenses in the smart grid: a survey. IET Cyber-Phys. Syst. Theory Appl. **1** (2016). https://doi.org/10.1049/iet-cps.2016.0019
23. V.K. Ponnusamy, P. Kasinathan, R. Madurai Elavarasan, V. Ramanathan, R.K. Anandan, U. Subramaniam, A. Ghosh, E. Hossain, A comprehensive review on sustainable aspects of big data analytics for the smart grid. Sustainability **13**, 13322 (2021). https://doi.org/10.3390/su132313322
24. O. Butt, M. Zulqarnain, T. Butt, Recent advancement in smart grid technology: future prospects in the electrical power network. Ain Shams Eng. J. **12** (2020). https://doi.org/10.1016/j.asej.2020.05.004
25. X. Niu, J. Li, J. Sun, K. Tomsovic, Dynamic detection of false data injection attack in smart grid using deep learning, in *2019 IEEE Power & Energy Society Innovative Smart Grid Technologies Conference (ISGT)* (2019)
26. S. Khan, K. Kifayat, A. Bashir, A. Gurtov, M. Hassan, Intelligent intrusion detection system in smart grid using computational intelligence and machine learning. Trans. Emerg. Telecommun. Technol. **32** (2021). https://doi.org/10.1002/ett.4062
27. K. Zhang, Z. Hu, Y. Zhan, X. Wang, K. Guo, A smart grid AMI intrusion detection strategy based on extreme learning machine. Energies **13**, 4907 (2020). https://doi.org/10.3390/en13184907
28. F. Chollet, Deep Learning with Python, 2nd edn (Manning Publications, 2021)
29. R. Hecht-Nielsen, Theory of the backpropagation neural network, in *Neural Networks for Perception*, vol. 2, pp. 65–93
30. F. Calero, C.A. Cañizares, K. Bhattacharya, Aggregated BESS dynamic models for active distribution network studies. IEEE Trans. Smart Grid **12**(3), 2077–2088 (2021). https://doi.org/10.1109/TSG.2020.3048648
31. K. Bhanot, S.K. Peddoju, T. Bhardwaj, A model to find optimal percentage of training and testing data for efficient ECG analysis using neural network. Int. J. Syst. Assur. Eng. Manag. **9**, 12–17 (2018). https://doi.org/10.1007/s13198-015-0398-7
32. T. Bhardwaj, H. Upadhyay, L. Lagos, Deep learning-based cyber security solutions for smart-city: application and review, in *Artificial Intelligence in Industrial Applications. Learning and Analytics in Intelligent Systems*, ed. by S.L. Fernandes, T.K. Sharma, vol. 25. (Springer, Cham, 2022). https://doi.org/10.1007/978-3-030-85383-9_12
33. M. Tarafdar Hagh, A. Mashhadi Kashtiban, Application of neural networks in power systems: a review, in *Proceedings of World Academy of Science, Engineering and Technology*, vol. 6 (2005)
34. T. Bhardwaj, R. Mittal, H. Upadhyay, L. Lagos, Applications of swarm intelligent and deep learning algorithms for image-based cancer recognition, in *Artificial Intelligence in Healthcare. Advanced Technologies and Societal Change* ed. by L. Garg, S. Basterrech, C. Banerjee, T.K. Sharma (Springer, Singapore, 2022). https://doi.org/10.1007/978-981-16-6265-2_9
35. G. Rolim, J.G. Zurn, Interpretation of remote backup protection for fault section estimation by a fuzzy exper system, in *IEEE PowerTech Conference* (2003), pp. 312–315
36. R. Lukomski, K. Wilkosz, Power system topology verification using artificial neural network utilization of measurement data, in *IEEE PowerTech Conference* (2003), pp. 180–186
37. S. Singh, S. Hussain, M.A. Bazaz, Short term load forecasting using artificial neural network, in *2017 Fourth International Conference on Image Information Processing (ICIIP)* (2017), pp. 1–5. https://doi.org/10.1109/ICIIP.2017.8313703

38. A.H. Hutama, S. Akbar, M.Z. Catur Candra, Medium term power load forecasting for Java and Bali power system using artificial neural network and SARIMAX, in *2018 5th International Conference on Data and Software Engineering (ICoDSE)* (2018), pp. 1–6. https://doi.org/10.1109/ICODSE.2018.8705837
39. Q. Jiang, J.-X. Zhu, M. Li, H.-Y. Qing, Electricity power load forecast via long short-term memory recurrent neural networks, in *2018 4th Annual International Conference on Network and Information Systems for Computers (ICNISC)* (2018), pp. 265–268. https://doi.org/10.1109/ICNISC.2018.00060
40. S. Panta, S. Premrudeepreechacharn, S. Nuchprayoon, C. Dechthummarong, S. Janjommanit, S. Yachiangkam, Optimal economic dispatch for power generation using artificial neural network, in *2007 International Power Engineering Conference (IPEC 2007)* (2007), pp. 1343–1348
41. K. Xu, Fault diagnosis method of power system based on neural network, in *2018 International Conference on Virtual Reality and Intelligent Systems (ICVRIS)* (2018), pp. 172–175. https://doi.org/10.1109/ICVRIS.2018.00049
42. A. Anika, M. Junaed-Al-Hossain, S. Hasibul Alam, Nahid-Al-Masood, Fault detection and classification of power system busbar using artificial neural network, in *2019 IEEE International Conference on Power, Electrical, and Electronics and Industrial Applications (PEEIACON)* (2019), pp. 31–34. https://doi.org/10.1109/PEEIACON48840.2019.9071944
43. Y. Zhang, Z. Rao, Research on information security evaluation based on artificial neural network, in *2020 3rd International Conference on Advanced Electronic Materials, Computers and Software Engineering (AEMCSE)* (2020), pp. 424–428. https://doi.org/10.1109/AEMCSE50948.2020.00098
44. A.N. Al-Masri, M.Z.A. Ab Kadir, A.S. Al-Ogaili, Y. Hoon, Development of adaptive artificial neural network security assessment schema for Malaysian power grids. IEEE Access **7**, 180093–180105 (2019). https://doi.org/10.1109/ACCESS.2019.2957884
45. R. Yan, G. Geng, Q. Jiang, Y. Li, Fast transient stability batch assessment using cascaded convolutional neural networks, in *2020 IEEE Power & Energy Society General Meeting (PESGM)* (2020), pp. 1–1. https://doi.org/10.1109/PESGM41954.2020.9282095
46. Y. Xu et al., Assessing short-term voltage stability of electric power systems by a hierarchical intelligent system. IEEE Trans. Neural Netw. Learn. Syst. **27**(8), 1686–1696 (2016). https://doi.org/10.1109/TNNLS.2015.2441706
47. S.K. Tiwary, J. Pal, ANN application for voltage security assessment of a large test bus system: a case study on IEEE 57 bus system, in *2017 6th International Conference on Computer Applications in Electrical Engineering-Recent Advances (CERA)* (2017), pp. 332–334. https://doi.org/10.1109/CERA.2017.8343350
48. T. Bhardwaj, S. Sharma, Fuzzy logic-based elasticity controller for autonomic resource provisioning in parallel scientific applications: a cloud computing perspective. Comput. Electr. Eng. **70**, 1049–1073 (2018). https://doi.org/10.1016/j.compeleceng.2018.02.050
49. K. Kadarla, S.C. Sharma, T. Bhardwaj, A. Chaudhary, A simulation study of response times in cloud environment for IoT-based healthcare workloads, in *2017 IEEE 14th International Conference on Mobile Ad Hoc and Sensor Systems (MASS)* (2017), pp. 678–683. https://doi.org/10.1109/MASS.2017.65
50. T. Bhardwaj, T.K. Sharma, M.R. Pandit, Social Engineering prevention by detecting malicious URLs using artificial bee colony algorithm, in *Proceedings of the Third International Conference on Soft Computing for Problem Solving*, ed. by M. Pant, K. Deep, A. Nagar, J. Bansal. Advances in Intelligent Systems and Computing, vol. 258 (Springer, New Delhi, 2014). https://doi.org/10.1007/978-81-322-1771-8_31.
51. P. Breviglieri, T. Erdem, S. Eken, Predicting smart grid stability with optimized deep models. SN Comput. Sci. **2**, 73 (2021). https://doi.org/10.1007/s42979-021-00463-5
52. T. Bhardwaj, S.C. Sharma, An autonomic resource provisioning framework for efficient data collection in cloudlet-enabled wireless body area networks: a fuzzy-based proactive approach. Soft Comput. **23**, 10361–10383 (2019). https://doi.org/10.1007/s00500-018-3587-x
53. T. Bhardwaj, H. Upadhyay, S.C. Sharma, Framework for quality ranking of components in cloud computing: regressive rank, in *2020 10th International Conference on Cloud Computing, Data*

Science & Engineering (Confluence) (2020), pp. 598–604. https://doi.org/10.1109/Confluence47617.2020.9058016
54. T.K. Sharma, M. Pant, T. Bhardwaj, PSO ingrained Artificial Bee Colony algorithm for solving continuous optimization problems, in *2011 IEEE International Conference on Computer Applications and Industrial Electronics (ICCAIE)* (2011), pp. 108–112. https://doi.org/10.1109/ICCAIE.2011.6162114
55. R. Vinayakumar, M. Alazab, K.P. Soman, P. Poornachandran, A. Al-Nemrat, S. Venkatraman, Deep learning approach for intelligent intrusion detection system. IEEE Access **7**, 41525–41550 (2019). https://doi.org/10.1109/ACCESS.2019.2895334
56. M. Tavallaee, N. Stakhanova, A.A. Ghorabani, Toward credible evaluation of anamaly-based intrusion-detection method. IEEE Trans. Syst. Man Cybern. Part C Appl. Rev. **40**(5), 516–524 (2010)
57. [Online] Available: https://www.ll.mit.edu/r-d/datasets/1998-darpa-intrusion-detection-evaluation-dataset
58. [Online] Available: https://www.ll.mit.edu/r-d/datasets/1999-darpa-intrusion-detection-evaluation-dataset
59. [Online] Available: https://www.ll.mit.edu/r-d/datasets/2000-darpa-intrusion-detection-scenario-specific-datasets
60. KDDCUP 1999 Dataset [online] available: http://kdd.ics.uci.edu/databases/kddcup99/kddcup99 Accessed 11/30/2021
61. NSL-KDD dataset [online] available: http://nsl.cs.unb.ca/nsl-kdd/. Accessed 11/30/2021
62. M. Tavallaee, E. Bagheri, W. Lu, A.A. Ghorbani, A detailed analysis of the KDD CUP 99 data set, in *2009 IEEE Symposium on Computational Intelligence for Security and Defense Applications* (2009), pp. 1–6.https://doi.org/10.1109/CISDA.2009.5356528
63. T. Bhardwaj, H. Upadhyay, S.C. Sharma, Autonomic resource provisioning framework for service-based cloud applications: a queuing-model based approach, in *2020 10th International Conference on Cloud Computing, Data Science & Engineering (Confluence)* (2020), pp. 605–610. https://doi.org/10.1109/Confluence47617.2020.9058266
64. T. Bhardwaj, H. Upadhyay, S.C. Sharma, Autonomic resource allocation mechanism for service-based cloud applications, in *2019 International Conference on Computing, Communication, and Intelligent Systems (ICCCIS)* (2019), pp. 183–187. https://doi.org/10.1109/ICCCIS48478.2019.8974515.
65. P. Gangwani, A. Perez-Pons, T. Bhardwaj, H. Upadhyay, S. Joshi, L. Lagos, Securing environmental IoT data using masked authentication messaging protocol in a DAG-based blockchain: IOTA tangle. Future Internet **13**, 312 (2021). https://doi.org/10.3390/fi13120312

Chapter 8
Learning-Based Model for Phishing Attack Detection

Jayesh Soni, Surya Sirigineedi, Krishna Sai Vutukuru, S. S. ChandanaEswari Sirigineedi, Nagarajan Prabakar, and Himanshu Upadhyay

Abstract The growing amount of phishing outbreaks is one of the main concern for security scientists nowadays. Signature-based approaches are used in the traditional methods for categorizing phishing websites. Such tools and techniques fail with advanced and complex phishing webpage. Consequently, learning-based algorithms are widely adopted in many industries. Such algorithms can be applied to analyze phishing websites. The detection accuracy can be increased with a large dataset and complex features. This chapter addresses the problem of analyzing such phishing activities on web pages. Explicitly, we propose a machine learning-based framework with various supervised learning algorithms such as Random Forest, Support Vector Machine, and Decision Trees. We perform the hyperparameter optimization of these algorithms using sci-kit-learn machine learning frameworks. In the end, we discussed the applied implementation of classifying the phishing attack on the real-world dataset available on Kaggle.

J. Soni (✉) · N. Prabakar
Knight Foundation School of Computing and Information Sciences, Florida International University, Miami, FL, USA
e-mail: jsoni@fiu.edu

N. Prabakar
e-mail: prabakar@cis.fiu.edu

H. Upadhyay
Applied Research Center, Florida International University, Miami, FL, USA
e-mail: upadhyay@fiu.edu

S. Sirigineedi
Department of Electrical and Computer Engineering, Florida International University, Miami, FL, USA
e-mail: ssiri005@fiu.edu

K. S. Vutukuru
Thornton Tomasetti, Miami, FL, USA
e-mail: kvutu001@fiu.edu

S. S. C. Sirigineedi
University Of Central Missouri, Warrensburg, USA

T. Bhardwaj et al. (eds.), *Artificial Intelligence in Cyber Security: Theories and Applications*, Intelligent Systems Reference Library 240,
https://doi.org/10.1007/978-3-031-28581-3_11

Keywords Phishing webpages · Machine learning · Random forest · Support vector machine · Decision trees

8.1 Introduction

National Institute of Science and Technology (NIST) describes cybersecurity as the capability to defend and shield the use of cyberspace from cyber attacks [1]. IBM defines cybersecurity/IT security as the practice of safeguarding sensitive information and critical systems from digital attacks [2]. Broadly, cybersecurity refers to identifying the threats and providing specific measures to protect the network systems and applications from within an organization threat and/or external threats. According to a recent IBM report, the annual global cost of the data breach was $4.24 million [3], with an increase of 10% from 2020. An exciting factor for 2021 data breach statistics was that remote working and digital transformation caused $1.07 million in losses in breaches. These costs include discovering and responding to the breach, downtime cost and revenue lost, and the damage caused to a business's reputation.

Typical kinds of such attacks include phishing attacks and ransomware attacks [4]. Phishing is a type of typical engineering attack often used to steal client records, including but not limited to usernames and passwords (login credentials) and confidential information such as credit card social security numbers. Such attacks occur when a cybercriminal, masquerades as a trusted entity, deceives a client/consumer into opening an instant message, text message, or most commonly, email. Often, in-house training to all the employees using the network systems in a firm could help prevent such attacks and protect IT security.

A blacklist developed by security experts is a set of websites that are unsafe or risky. Phish Tank comprises an extensive list of such websites. Google Safe Browsing provided by Google can help users to some extent. The study [5] developed an approach for creating an efficient blacklist of such phishing sites. Their approach distinguishes a website as normal or phishing in three steps. First, they checked the commands, whether they were benign or malicious. Next, they verify if the content of the website contains any urgency text which is objectively conjoint in malicious attacks. Lastly, they perform the link analysis to check the soundness of the associated link. With the rise of creating websites using the available framework and automated scripts, such framework misses the malicious link. Learning based algorithms are widely employed recently with the advent of big data and high computational power [6–10].

The rest of the chapter is summarized as follows: Sect. 2 gives an in-depth overview of learning-based algorithms. Section 3 discusses the applications of machine learning in the cybersecurity domain. Section 4 discusses the open-source framework and library. Section 5 describes these algorithms' practical real-time use cases to solve the phishing activity detection problem. Finally, we conclude in Sect. 6 with future work discussion.

8.2 Learning-Based Algorithms

Machine learning (ML) is a sub-field of artificial intelligence that deals with the research procedures that offer systems, the capacity to study and progress from knowledge [11–15] inevitably. Typically, such algorithms consist of identifying underlying patterns from data to make accurate decisions. Hence, typically the amount of data used in training such models would be proportional to prediction accuracy. Depending on the types of datasets used, output requirements, the problem type to be solved, and the underlying methodology, ML techniques can be categorized into supervised, unsupervised, and semi-supervised learnings.

8.2.1 Supervised Machine Learning

Supervised Machine Learning is typically used when the precise definition of input variables and output targets are known. In a way, this algorithm tries to map the given input variables to the expected output variables and creates a mapping function that could predict the output based on the provided input. These methods can be broadly classified into regression and classification [16, 17]. Regression is typically used when the output is a numerical value, and classification is generally used when the output falls in a known category of outputs. Below we discuss various popular classification-based machine learning algorithms.

8.2.1.1 Support Vector Machine

The Support Vector Machine (SVM) [18] technique is a straightforward but effective supervised machine learning algorithm for developing regression and classification models. SVM is effective with both linearly and non-linearly separable datasets. Even with a small amount of data, the support vector machine technique performs admirably (Fig. 8.1).

There are two types of Support Vector Machines:

Linear SVM: For data that can be separated linearly, linear SVM is employed. A dataset is linearly separable if it can be categorized into two categories using only a single straight line. It is widely used to address problems involving linear regression and classification.

$$G(m, m_j) = OP(m, m_j) \tag{8.1}$$

Here m, m_j embodies the data.

Nonlinear SVM: Nonlinear SVM is used to classify data that is not linearly separated and be characterized using a line that is straight. In this case, it is referred to as a nonlinear SVM classifier since more features can be added to fit a hyperplane

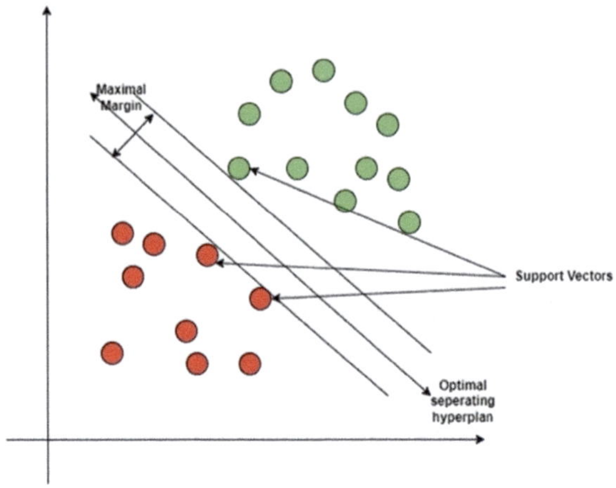

Fig. 8.1 Support vector machine

rather than a two-dimensional space, enabling greater nonlinear data flexibility.

$$G(m, m_j) = \exp(-\gamma||m - m_j||)^2 \tag{8.2}$$

Here ɣ contrasts from 0 to 1.

Advantage

1. SVM uses minimal memory.
2. SVM is more compelling in high dimensions.

Disadvantage

1. Not suitable for large datasets.
2. SVM model gives bad results with unbalanced data sets.

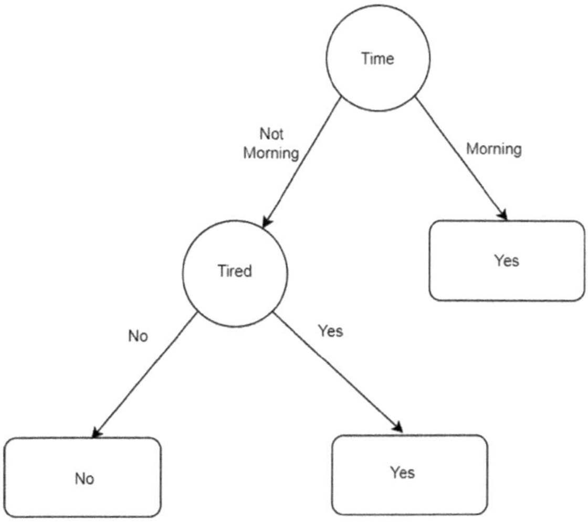

Fig. 8.2 Decision tree

8.2.1.2 Decision Tree

Most mathematical models fail when there is a nonlinearity. Thus, to overcome this problem, scientist has employed a decision tree [19–21]. Moreover, random forest machine learning methods heavily rely on decision tree algorithms. This algorithm can work with nominal and numerical features. Figure 8.2 is a pictorial representation of a decision tree where it describes a person who will drink a coffee or not based on different features like time of the day and tiredness. Based on this, it is easy to predict the person's chances of looking for a coffee.

Advantage

1. They have better results with nonlinear datasets.
2. It is simpler to interpret the decision tree.

Disadvantage

1. High changes of the model being overfitted.
2. Training model becomes difficult if levels increase.

8.2.1.3 Random Forest

Random forest [22–24] is an ensemble of decision trees. Each decision tree is built independently using random subsets generated from the input dataset, and each tree casts a unit vote—the class with the most votes to classify an input (Fig. 8.3).

Advantages

1. Random forest is a diverse model as it can be used for regression and classification problems.

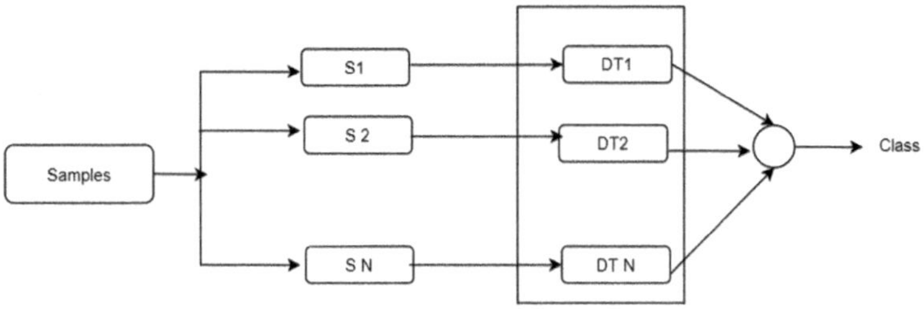

Fig. 8.3 Random forest

2. This is fast to train and highly unlikely to get overfit.
3. Random forest algorithm provides better results where classes are imbalanced in datasets.
4. Resources used for the training model are less.

Disadvantages

1. It is a kind of black box that is difficult to investigate on the model.
2. It is inefficient to use for practical purposes as it takes more time to generate the results.

8.2.1.4 Logistic Regression

The logistic regression model [25] is employed to overcome the short comes in the linear regression model. Unlike linear regression, where data is fitted into a hyperplane in logistic regression, the data is squeezed into a mathematical model between 0 and 1. The logistic regressions are used to solve the binary dependent variables like a person having a tumor or not, chances of raining. Equation 8.3 is the mathematical representation of the logistic regression where beta is coefficient and X is independent variables. Moreover, y is the estimated probability of being a category (Fig. 8.4).

$$\ln \frac{y}{1-y} = \beta_0 + \beta_1 X_1 + \beta_2 X_2 \ldots \beta_N X_N \tag{8.3}$$

Advantage

1. It can extend to multinomial classes.
2. Regularization can be used to remove overtraining of a logistic regression model.

Disadvantage

1. Overfitting is an issue.
2. It has a considerable disadvantage with higher dimensional features.

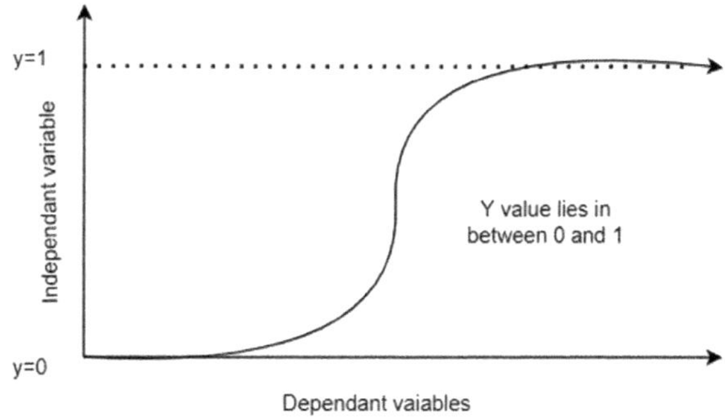

Fig. 8.4 Logistic regression

8.2.2 Unsupervised Machine Learning

Unsupervised machine learning is typically used when only the input information is known, and there is no corresponding output variable to map to it. They recognize the inherent patterns in the input to understand the characteristics of the data. Clustering is a common type of unsupervised ML technique in which inherent groups are identified in the data, and this is used to predict output for a different set of input data.

8.2.3 Semi-Supervised Machine Learning

It is an intermediate approach between supervised and unsupervised ML techniques. Such algorithms are trained using a mixture of categorized and uncategorized data. Typically, the data consists majority of uncategorized, with a small amount of categorized data [26, 27]. Such uncategorized data is first assembled using an unsupervised learning algorithm, and then categorized data is used to categorize the rest of the uncategorized data.

In today's world, every field, including science, industry, and management, has been producing large amounts of data making the interpretation, post-processing, and analysis cumbersome. To solve this problem, deep learning [28, 29] comes into the picture with a novel methodology combining a variety of learning algorithms instead of a single method to learn complex prediction models containing many hidden layers of neural networks [30, 31].

8.3 ML Applications in Cyber Security Domain

Machine learning is being used to detect malicious activity in the system and block them before they cause greater damage to the organization. Most of this machine learning-based system continuously monitors the different system metrics like network usage, CPU utilization, and RAM based on the signature usage of the application to considers them as malicious activity or benign activity. Latest antiviruses collect the system calls data and compare this sequence with the original call sequence, and based on the similarity percentage, if there is too much deviation, those sequences are considered malicious. There has been enormous research on detecting phishing webpage using natural language processing. These natural language processing techniques like lexical features, type setting, etc., are efficient in classifying whether a URL is a phishing attack or not.

Machine learning is also used in cyber security tasks, which are repetitive security tasks that involve low decision-making activities, and faster response is needed like ransomware attacks. When a cyber-attack happens on an organization, the security team must respond to the attack very quickly. Delay in counter measurements may result in substantial financial loss to the organization. Furthermore, cloud based services are widely adopted by multiple organizations [32–34]. Network Intrusion Detection System are rules-based or need human intervention in making decisions. Numerous NIDS systems have used deep learning-based models that can classify network packets without human intervention in their process. Most of the decisions here are solely made by machines. There are fewer errors during detection or decision-making with less response time.

8.4 Open Source Library and Evaluation Metrics

Tensorflow: Google developed Tensorflow as an open-source framework for deep learning algorithms [35].

Keras: Keras is a wrapper on Tensorflow and is used by many societies around the world. The code written in Keras is internally converted to TensorFlow for further execution. It has functional API (Application Programming Interface) and Sequential API [36].

Scikit-learn: Scikit-learn deals with various learning-based algorithms (both supervised and unsupervised) [37].

We used the following metrics to evaluate the model.

$$\text{Recall} = \text{TP}/(\text{TP} + \text{FN}) \tag{8.4}$$

$$\text{Precision} = \text{TP}/(\text{TP} + \text{FP}) \tag{8.5}$$

$$\text{Specificity} = \text{TN}/(\text{TN} + \text{FP}) \tag{8.6}$$

$$\text{Accuracy} = (\text{TP} + \text{TN})/(\text{TP} + \text{FP} + \text{FN} + \text{TN}) \tag{8.7}$$

$$\text{False Negative Rate (FNR)} = \text{FN}/(\text{TP} + \text{FN}) \tag{8.8}$$

$$\text{False Positive Rate (FPR)} = \text{FP}/(\text{FP} + \text{TN}) \tag{8.9}$$

where TP: True positive, FN: False Negative, TN: True Negative, and FP: False Posi-tive.

8.5 Practical Implementation

The proposed framework is described in Fig. 8.5. It has mainly three stages: Dataset Collection, Data Pre-Processing, Detection Algorithm Training, and the final evaluation. Below we provide details for each of the stages in brief.

Stage-1: Data Collection

We use the open-source Phishing Detection dataset available on Kaggle.There are 48 features mined from 5000 legitimate web pages and 5000 phishing webpages.

Stage-2: Data Preprocessing

We perform the following data preprocessing tasks:

(1) Normalize the dataset and remove the feature with zero variance.
(2) Split the dataset for training, validation, and testing using stratified sampling.

Stage-3: Algorithm Training

This is the primary step of the whole structure where we train multiple algorithms.
The following algorithms are trained and validated with the test data.

(1) Decision Trees
(2) Random Forest
(3) Support Vector Machine
(4) Logistic Regression

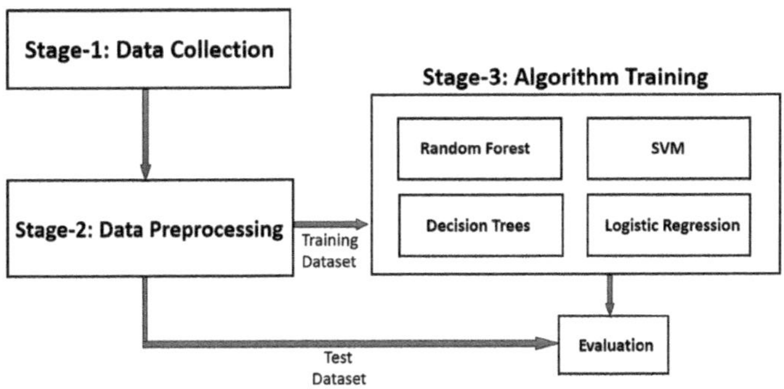

Fig. 8.5 Proposed framework

Table 8.1 Evaluated metrics

Algorithm	Accuracy	Precision	Recall	F1_score
Decision tree	0.968	0.970	0.971	0.971
Random forest	*0.975*	*0.974*	*0.969*	*0.972*
SVM	0.970	0.962	0.951	0.950
Logistic regression	0.931	0.947	0.941	0.949

From Table 8.1, we can depict that random forest distinguished the phishing website from the legitimate website with 97% accuracy.

8.6 Conclusion

Each distinct person connects to the internet for a specific purpose in the modern world. Cyber Attacks misleadusers into believing their false websites and lead themwith false actions to leak their personal information. Phishingattacks attempt to assemble users' data via emails or web links without their permission. This chapterdiscussed various machine learning-based algorithms with their advantages and disadvantages. Furthermore, we gave a practical implementation of applying these algorithms to solving the phishing attack detection problem using a real-world open-source dataset.

References

1. Joint Task Force Transformation Initiative, Guide for conducting risk assessments, National Institute of Standards and Technology, Gaithersburg, MD, NIST SP 800–30r1 (2012)
2. What is Cybersecurity? | IBM. https://www.ibm.com/topics/cybersecurity. Accessed 26 Jan 2022
3. Cost of a Data Breach Report 2021, *Risk Quantif.*, p. 73
4. S.S. Sirigineedi, J. Soni, H. Upadhyay, Learning-based models to detect runtime phishing activities using urls. In *Proceedings of the 2020 the 4th International Conference on Compute and Data Analysis* (2020), pp. 102–106
5. J. Kang, D. Lee, Advanced white list approach for preventing access to phishing sites, in *2007 International Conference on Convergence Information Technology (ICCIT 2007)* (IEEE, 2007), pp. 491–496
6. J. Soni, N. Prabakar, Effective machine learning approach to detect groups of fake reviewers, in *Proceedings of the 14th International Conference on Data Science (ICDATA'18), Las Vegas, NV* (2018), pp. 3–9
7. J. Soni, N. Prabakar, J. H. Kim, Prediction of component failures of telepresence robot with temporal data, in *30th Florida Conference on Recent Advances in Robotics* (2017)
8. J. Soni, N. Prabakar, H. Upadhyay, Behavioral analysis of system call sequences using LSTM Seq-Seq, cosine similarity and Jaccard similarity for real-time anomaly detection, in *2019 International Conference on Computational Science and Computational Intelligence (CSCI)* (IEEE, 2019), pp. 214–219

9. G.S. Thejas, J. Soni, K. Chandna, S.S. Iyengar, N.R. Sunitha, N. Prabakar, Learning-based model to fight against fake like clicks on Instagram posts, in *2019 SoutheastCon* (IEEE, 2019), pp. 1–8
10. S.B. Kotsiantis, I.D. Zaharakis, P.E. Pintelas, Machine learning: a review of classification and combining techniques. Artif. Intell. Rev. **26**(3), 159–190 (2006). https://doi.org/10.1007/s10462-007-9052-3
11. T. Peng, I. Harris, Y. Sawa, Detecting phishing attacks using natural language processing and machine learning, in *2018 IEEE 12th International Conference on Semantic Computing (ICSC)* (IEEE, 2018), pp. 300–301
12. T. Bhardwaj, R. Mittal, H. Upadhyay, L. Lagos, Applications of swarm intelligent and deep learning algorithms for image-based cancer recognition, in *Artificial Intelligence in Healthcare* (Springer, Singapore, 2022), pp. 133–150
13. P. Gangwani, A. Perez-Pons, T. Bhardwaj, H. Upadhyay, S. Joshi, L. Lagos, Securing environmental IoT data using masked authentication messaging protocol in a DAG-based blockchain: IOTA tangle. Future Internet **13**(12), 312 (2021)
14. T. Bhardwaj, C. Reyes, H. Upadhyay, S.C. Sharma, L. Lagos, Cloudlet-enabled wireless body area networks (WBANs): a systematic review, architecture, and research directions for QoS improvement. Int. J. Syst. Assur. Eng. Manage. 1–25 (2021)
15. T. Bhardwaj, H. Upadhyay, S.C. Sharma, Framework for quality ranking of components in cloud computing: regressive rank, in *2020 10th International Conference on Cloud Computing, Data Science & Engineering (Confluence)* (IEEE, 2020), pp. 598–604
16. J. Soni, N. Prabakar, H. Upadhyay, Feature extraction through deepwalk on weighted graph, in *Proceedings of the 15th International Conference on Data Science (ICDATA'19), Las Vegas, NV* (2019)
17. T. Bhardwaj, H. Upadhyay, L. Lagos, Deep learning-based cyber security solutions for smart-city: application and review, in *Artificial Intelligence in Industrial Applications* (Springer, Cham, 2022), pp. 175–192
18. D.A. Pisner, D.M. Schnyer, Support vector machine, in *Machine Learning* (Academic Press, 2020), pp. 101–121
19. J. Su, H. Zhang, A fast decision tree learning algorithm, in *Aaai*, vol. 6 (2006), pp. 500–505
20. Y.Y. Song, Y. Lu, Decision tree methods: applications for classification and prediction. Shanghai Arch. Psychiatry **27**(2), 130–135 (2015). https://doi.org/10.11919/j.issn.1002-0829.215044
21. V. Verma, K.S. Vutukuru, S.S. Divvela, S.S. Sirigineedi, Internet of things and machine learning application for a remotely operated wetland siphon system during hurricanes, in *Water Resources Management and Sustainability* (Springer, Singapore, 2022), pp. 443–462
22. Y. Wang, R. Xiong, J. Wang, J. Zhang, Multi-class assembly parts recognition using composite feature and random forest for robot programming by demonstration. IEEE Int. Conf. Robot. Biomimetics (ROBIO) **2015**, 698–703 (2015). https://doi.org/10.1109/ROBIO.2015.7418850
23. T.M. Oshiro, P.S. Perez, J.A. Baranauskas, How many trees in a random forest? in *International Workshop on Machine Learning and Data Mining in Pattern Recognition* (Springer, Berlin, Heidelberg, 2012), pp. 154–168
24. W. Hu, M. Fey, M. Zitnik, Y. Dong, H. Ren, B. Liu, J. Leskovec, Open graph benchmark: datasets for machine learning on graphs. Adv. Neural. Inf. Process. Syst. **33**, 22118–22133 (2020)
25. D.G. Kleinbaum, K. Dietz, M. Gail, M. Klein, M. Klein, *Logistic Regression* (Springer-Verlag, New York, 2002), p.536
26. O. Chapelle, B. Scholkopf, A. Zien, Eds., Semi-Supervised Learning (Chapelle, O. et al., Eds.; 2006) [Book reviews]. IEEE Trans. Neural Netw. **20**(3), 542–542 (2009). https://doi.org/10.1109/TNN.2009.2015974
27. J. Soni, N. Prabakar, H. Upadhyay, Visualizing high-dimensional data using t-distributed stochastic neighbor embedding algorithm, in *Principles of Data Science* (Springer, Cham, 2020), pp. 189–206
28. J. Soni, S.K. Peddoju, N. Prabakar, H. Upadhyay, Comparative analysis of LSTM, one-class SVM, and PCA to monitor real-time malware threats using system call sequences and

virtual machine introspection, in *International Conference on Communication, Computing and Electronics Systems* (Springer, Singapore, 2021), pp. 113–127
29. J. Soni, N. Prabakar, H. Upadhyay, Comparative analysis of LSTM sequence-sequence and auto encoder for real-time anomaly detection using system call sequences (2019)
30. G.E. Hinton, R.R. Salakhutdinov, Reducing the dimensionality of data with neural networks. Science **313**(5786), 504–507 (2006)
31. Y. Bengio, A. Courville, P. Vincent, Representation learning: a review and new perspectives. IEEE Trans. Pattern Anal. Mach. Intell. **35**(8), 1798–1828 (2013). https://doi.org/10.1109/TPAMI.2013.50
32. T. Bhardwaj, H. Upadhyay, S.C. Sharma, Autonomic resource provisioning framework for service-based cloud applications: a queuing-model based approach, in *2020 10th International Conference on Cloud Computing, Data Science & Engineering (Confluence)* (IEEE, 2020), pp. 605–610
33. T. Bhardwaj, H. Upadhyay, S.C. Sharma, An autonomic resource allocation framework for service-based cloud applications: a proactive approach, in *Soft Computing: Theories and Applications* (Springer, Singapore, 2020), pp. 1045–1058
34. V. Verma, L. Bian, D. Ozecik, S.S. Sirigineedi, A. Leon, Internet-enabled remotely controlled architecture to release water from storage units, in *World Environmental and Water Resources Congress* (2021), pp. 586–592
35. J.V. Dillon, et al.,Tensorflow Distributions. arXiv preprint arXiv:1711.10604 (2017)
36. N. Ketkar, *Introduction to Keras. Deep Learning with Python* (Apress, Berkeley, CA, 2017), pp.97–111
37. F. Pedregosa, et al.,Scikit-learn: machine learning in Python. J. Mach. Learning Res. **12**, 2825–2830 (2011)

Chapter 9
Deep Learning Approach for Detection of Fraudulent Credit Card Transactions

Jayesh Soni, Pranav Gangwani, Surya Sirigineedi, Santosh Joshi, Nagarajan Prabakar, Himanshu Upadhyay, and Shrirang Ambaji Kulkarni

Abstract Instead of cash, people tend to use credit cards with the swift technological growth in the modern world. This unlocks the door for fraudulent individuals to utilize these cards in a wicked method. Every year, it costs billions of dollars in credit card transaction fraud to card issuers. There are no static patterns in fraud. Their behavior constantly changes. New technologies allow fraudsters to use the online medium and other techniques for implementing frauds. It is vital to learn the behavior patterns. The detection accuracy can be increased with a large dataset and complex features. This chapter addresses the problem of analyzing fraudulent credit card transactions. Explicitly, we propose a deep learning-based framework with various unsupervised learning algorithms. We perform the Hyperparameter optimization of these algorithms using sci-kit-learn machine learning frameworks and popular deep learning framework TensorFlow. In the end, we discussed the applied implementation of detecting the fraudulent transactions on the real-world dataset available on Kaggle.

J. Soni (✉) · P. Gangwani · S. Sirigineedi · S. Joshi · N. Prabakar · H. Upadhyay
Florida International University, Miami, FL, USA
e-mail: jsoni@fiu.edu

P. Gangwani
e-mail: pgang002@fiu.edu

S. Sirigineedi
e-mail: ssiri005@fiu.edu

S. Joshi
e-mail: sajoshi@fiu.edu

N. Prabakar
e-mail: prabakar@cis.fiu.edu

H. Upadhyay
e-mail: upadhyay@fiu.edu

S. A. Kulkarni
University of Central Florida, Orlando, USA
e-mail: sakulkarni@ucf.edu

T. Bhardwaj et al. (eds.), *Artificial Intelligence in Cyber Security: Theories and Applications*, Intelligent Systems Reference Library 240,
https://doi.org/10.1007/978-3-031-28581-3_13

Keywords Credit card · Fraud transactions · Deep learning · AutoEncoder · Anomaly detection

9.1 Introduction

Cybersecurity application ensures the protection of organizations' hardware, software networks and the information stored in the computer systems from any kind of viruses, damage, unauthorized access, or any malicious activity on the systems that could bring in a substantial amount of data damage, which could lead to breaking of the confidentiality, integrity, and availability of the organization. Cybersecurity has become the need of the hour, with cybercrimes rates increasing by 45% from 3.3 million in 2019 to 4.8 million in 2020 [1].

The knowledge and awareness of cyber security is essential for all the internet users to perform daily activities. As acknowledged by the specialists in the security industry, people need to be educated in the cyber security aspects and be more cyber secure. To address different cyber security challenges faced by individuals, a "self-renewing" standpoint is required. In an organization this is accomplished by adopting information security culture (ISC), similarly cyber security culture (CSC) needs to be taken up in societal circumstances. This provides a different viewpoint to think about what CSC should consist of and how CSC differs from the ISC. Some of the different security aspects that need to be in place to be cyber-secure include: Network Security, an activity to secure the computer systems and networks from the assailant and intruders. Operational Security, the procedures in protecting the data assets and handling the permissions to access the data. Application security, process of securing the software applications to make sure they are free from threats by the timely updates and testing. Information security is a procedure to manage the privacy and integrity of the data during the transfer and the storage. A statistical method for fraud detection generates a specific score for a particular transaction. The value of the threshold is always a practical challenge. Artificial neural networks (ANN) [2, 3], random forest [4], decision trees [5], hidden Markov models [6] are widely studied in the research community for credit card fraud detection. Such learning-based algorithms have been widely employed recently with the advent of big data and high computational power [7–9]. To detect unusual behavior, unsupervised methods group customer characteristics into various clusters. When a new behavior shows a deviation beyond the threshold from the expected behavior, it is labeled as fraud. Self-organizing maps [10] and peer group analysis [11] are unsupervised methods. Supervised learning needs lots of suitable labeled datasets. However, in real-life scenarios, some of the credit card transactions may be mislabeled. Furthermore, such a dataset is highly imbalanced. Therefore, this chapter discusses unsupervised approaches for detecting fraudulent transactions. Furthermore, the data can be processed using cloud computing techniques [12–14]. The rest of the chapter is summarized as follows. Section 9.2 gives an in-depth overview of learning-based algorithms. Section 9.3 provides an understanding of various anomaly detection algorithms. Section 9.4 discusses the applications of learning-based algorithms in Anomaly Detection. Section 9.5 describes these algorithms' practical real-time use cases to solve the credit card fraud detection problem. Finally, we conclude in Sect. 9.6.

9.2 Learning-Based Algorithms

Machine learning: It is a data driven approach to train computers to think and act like humans perform the activities and improve their knowledge by autonomous learning over a period of time. This is achieved by feeding the information in terms of data points and real world interactions. Machine learning algorithms are classified as supervised, unsupervised, and semi-supervised.

9.2.1 Supervised Machine Learning

Supervised machine learning algorithms can predict the future events in the new data by learning from the labeled examples. They use the training dataset consisting of inputs and correct output data points. The trained model can predict new input after sufficient training with all the possible scenarios. The accuracy of the prediction is calculated using the loss function and retrained with additional samples and modifying the hyper parameters to minimize the error. Supervised learning can address two types of problem sets—classification and regression.

9.2.2 Unsupervised Machine Learning

Unsupervised machine learning algorithms are used to analyze the datasets which are not labeled. This type of algorithmlearns the hidden pattern in the unlabeled data. These algorithms are ideal for exploratory data analysis, image recognition etc. Clustering is one of the commonly used data mining algorithms to segregate the unlabeled data into groups based on the differences and the similarities. Anomaly detection is another broadly used unsupervised learning methodology which detects the data points or outliers, which are rare events far from the reality of the normal activity.

9.2.3 Semi-Supervised Machine Learning

The Semi-supervised machine learning is a technique used when the dataset contains a small number of labeled data and a huge number of unlabeled data. This type of the problem sets are challenging as it is difficult to implement supervised or unsupervised algorithms, hence semi-supervised algorithms are required. Semi-supervised learning is a combination of unsupervised and supervised methods, and is very useful when it is challenging to collect the large amount of labeled data. The semi-supervised learning algorithms work on the concept of pseudo labeling, meaning model training is performed with the minimal labeled training data. The main advantage of the semi-supervised learning is it extends the dependent information provided by the labeled samples to a larger amount of unlabeled dataset with the minimal accuracy loss.

Deep Learning: It is a learning technique based on the concept of artificial neural networks to analyze the complex data patterns to create an efficient and accurate predictive model. It is developed by mimicking the structure and functionality of the neurons in the human brain. Multiple neural layers are employed together to form an extensive deep neural network (DNN), which is the core computing layer of the deep learning algorithm. DNN uses many possible examples in the training phase to learn the relationship between inputs and outputs. This method automates the feature identification and extraction, hence reducing the manual work of feature identification. This leads to higher accuracy in prediction results. It is fast and well suited for the dataset containing large amounts of data points.

We restrict our discussion to anomaly detection in this chapter.

9.3 Anomaly Detection Algorithms

Anomaly Detection refers to detecting unexpected events embedded in the enormous amount of canonical data. The tasks include training a model with examples from one class of data. In machine learning, we use the class (normal and abnormal) of data to train the algorithm [15–17]. This model is used to classify test samples that may contain new classes of samples. It identifies new profiles (concepts) of untagged data that may indicate the occurrence of new concepts, changes in general concepts, or the presence of noise. The discovery of new concepts has intrigued the knowledge discovery community, commonly referred to as novelty detection or open set detection. Majority of such algorithms operate in two stages. The first stage is completed offline. Algorithms learn a characteristic concept for each class from a large dataset. The current decision model analyzes each unlabeled example in the channel.

9.3.1 Machine Learning-Based Anomaly Detection Algorithms

We discuss the following machine learning-based anomaly detection algorithms in detail.

1. One-Class Support Vector Machine (OCSVM)
2. Isolation Forest
3. Elliptical Envelope
4. Mahalanobis Distance Metric

9.3.1.1 One-Class Support Vector Machine (OCSVM)

Support Vector Machine has been a general approach for nonlinear classification problems. One Class Support Vector Machine [18] is an altered form of support vector machine with only one class of dataset. It is a widespread anomaly detection and outlier detection technique. This algorithm maps each data point to a higher dimensional feature space and then uses a trailing sphere so that most data falls inside it. By linearizing the data, it gives better results with non-monotone and non-linearly separable data.

9.3.1.2 Isolation Forest

The algorithm's primary function is to "isolate" anomalies by constructing decision trees over random features [19]. Because random partitioning creates significantly shorter paths for anomalies. A decision tree is the basic building block for the isolation forest. These decision trees are developed using information like the Gini index and entropy. Dataset is randomly divided into split values based on attribute values, and the process is repeated accordingly till a tree is formed. The samples on a shorter branch of the tree are anomalies. It is less prone to errors as they are built using isolation techniques rather than k- means and takes less time to train the model.

9.3.1.3 Elliptical Envelope

The elliptical envelope is logically built on the hypothesis that data comes from a known distribution [20]. So, if we take an ellipse around a Gaussian distribution of the data, everything outside the ellipse is considered an outlier. The data points lie inside this ellipse will be considered inliers, and points lying outside are considered outliers.

9.3.1.4 Mahalanobis Distance Metric

The Mahalanobis distance [21] is a distance unit of measure that determines the gap among a point and its distribution. It performs excellently on a multivariate dataset since it calculates covariance between variables to determine the distance between two points. In other words, Mahalanobis computes the distance between points "P1" and "P2" by taking standard deviation into account.

9.3.2 Deep Learning-Based Anomaly Detection Algorithms

We discuss the following deep learning-based anomaly detection algorithms in detail.

1. Long Short Term Memory (LSTM)
2. Bidirectional LSTM (Bi-LSTM)
3. Gated Recurrent Unit (GRU)
4. AutoEncoder

9.3.2.1 Long Short Term Memory (LSTM)

An unsupervised deep learning algorithm that falls in Recurrent Neural Networks (RNN) is called a Long-Short-Term Memory (LSTM). This algorithm contains LSTM blocks, also known as memory cell units [22–25]. LSTM can remember state values for an indefinite period due to these memory cell units. There are three separate gate units contained within an LSTM block, and these gate units can learn when to maintain, use, or delete a state. Social behavior analysis, natural language processing (NLP), and stock forecasting are some of the few examples of applications where LSTM has been used effectively.

In order to build an LSTM model, the first stage should involve deciding what information should be discarded from cell states, and this conclusion is completed by a layer known as "forget gate" or a "sigmoid layer". The model of LSTM will check m_{t-1} and y_t according to Eq. (9.1), and for each number in the cell state, S_{t-1}, the algorithm will produce a value between 0 and 1. Output 1 means "completely keep this state" and output 0 means "completely discard this state".

$$g_t = \sigma\left(w_F\left[m_{t-1}, y_t\right] + b_F\right) \tag{9.1}$$

Next, a decision must be made about which data must be saved in the cell state. The input gate layer, a sigmoid layer, chooses which values will be updated first, as shown in Eq. (9.2). The state is then updated with a $tanh$ layer that generates a vector of new candidate values, S'_t as shown in Eq. (9.3). To update the state of the old cell S_{t-1} into a new cell state, S_t; these two steps are combined.

$$l_t = \sigma\left(w_i\left[m_{t-1}, y_t\right] + b_i\right) \tag{9.2}$$

$$S'_t = \sigma\left(w_c\left[m_{t-1}, y_t\right] + b_c\right) \quad (9.3)$$

The next step involves the multiplication of the old state by f_t which is the forgetting state's ratio. Then, to scale every state value $l_t \cdot S'_t$ must be added as shown in Eq. (9.4).

$$S_t = g_t \cdot S_{t1} + l_t \cdot S'_t \quad (9.4)$$

Lastly, the result is computed using the cell states that have been further filtered. According to Eq. (9.5), the cell states to output are determined by a "sigmoid layer", and to convert the values in the range of -1 to 1, the filtered states of the cell are then passed through tanh. The multiplication of the sigmoid gate output with the tanh value is the resulting output.

$$O_t = \sigma\left(w_o\left[m_{t-1}, y_t\right] + b_o\right) \quad (9.5)$$

$$m_t = O_t \cdot tanh\ tanh(S_t) \quad (9.6)$$

9.3.2.2 Bidirectional LSTM

A neural network that can store sequence data in both directions is called bidirectional LSTM (bi-LSTM). The two mentioned directions can be forward or backward [26].

The feature that differentiates bi-LSTM from the normal LSTM is its bidirectional nature, allowing the input to flow in two directions. The input only flows in one direction, i.e., either forward or backward, in regular LSTM. On the other hand, in bi-LSTM, we can tune it to make the input flow in both directions, preserving the past and future data.

Figure 9.1 shows an example of a bidirectional LSTM. In this figure, the information flow can be seen from the forward and backward layers. This type of bidirectional architecture is commonly used in tasks where the sequence-to-sequence flow is required. Bidirectional LSTM networks are widely utilized in forecasting models, audio recognition, and text classification.

9.3.2.3 Gated Recurrent Unit (GRU)

Gated Recurrent Unit (GRU) is a type of RNN and a gating mechanism introduced in 2014. GRU is similar to an LSTM, and it contains an additional gate called forget gate [27]. GRU also has fewer parameters since an output gate is missing. The performance of GRU on specific applications such as NLP, "speech signal modeling," and polyphonic music modeling was proven to be comparable to that of LSTM in some cases. On some smaller and less frequent datasets, GRUs have shown excellent

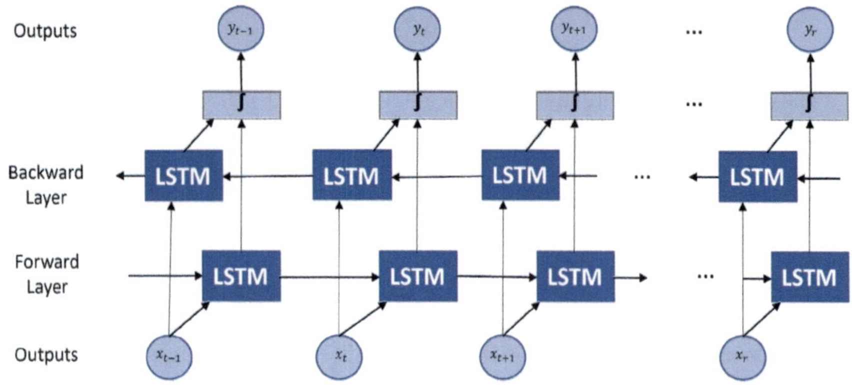

Fig. 9.1 Bidirectional—LSTM neural network

performance. The final model is easier to understand than ordinary LSTM models, and it is now gaining popularity.

The information in GRU is modulated within the unit similar to an LSTM but without using a separate unit of a memory cell. At time t, the GRU activation h_t^j is a linear interpolation of the previous activation h_{t-1}^j and the candidate activation, h'^j_t:

$$h_t^j = \left(1 - z_t^j\right)h_t^j + z_t^j h_t^{'j} \tag{9.7}$$

The update gate z_t^j controls the amount of the unit's activation that is updated –

$$z_t^j = \sigma(W_z x_t + U_z h_{t-1})^j \tag{9.8}$$

The computation of the candidate activation function h'^j_t is done like the "update gate":

$$h_t^{'j} = tanh(Wx_t + U(r_t . * h_{t-1}))^j \tag{9.9}$$

where r_t signifies a group of reset gates and ". *" means elementwise multiplication. When $r_t == 0$, that means the "reset gate" is turned off, the unit may not recall the past. Allowing the unit to read the first symbol in an input sequence is equivalent. The computation of the reset gate can be done by using the following equation.

$$r_t^j = \sigma(W_r x_r + U_r h_{t-1})^j \tag{9.10}$$

The "update gate" z determines the importance of the previous state. The units with short-term dependencies will contain the active reset gates r. On the other hand, the units with long-term dependencies contain the active update gates z.

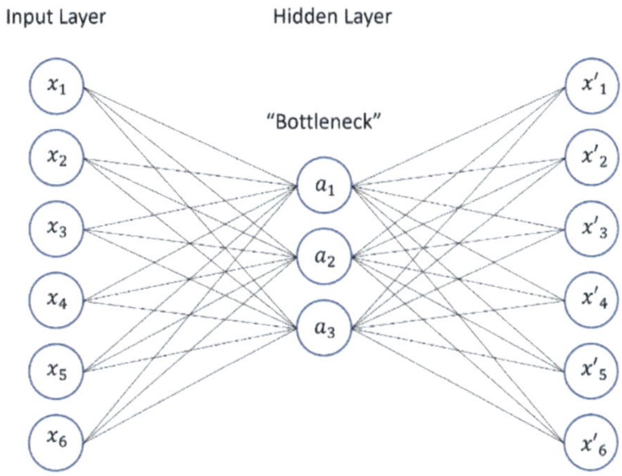

Fig. 9.2 AutoEncoder network

9.3.2.4 AutoEncoders

An AutoEncoder is used to learn unsupervised data encodings and is a form of an artificial neural network [28]. The goal of an AutoEncoder is to train the network to capture the essential bits of the input picture to learn a lower-dimensional representation (encoding) for higher-dimensional data, often for dimensionality reduction. The deep walk is another technique to extract the feature vector [29]. Figure 9.2 shows the architecture of an AutoEncoder.

There are three parts in AutoEncoder architecture as mentioned below:

Encoder: The encoder is a module that provides an encoded version of the input data that is often several orders of magnitude less than the original data.

Bottleneck: The compressed knowledge representations are stored in this module, making it the most critical portion of the network.

Decoder: The decoder is a module that assists the network in "decompressing" knowledge representations and reassembling data from its encoded state. The output is then compared to a reference point.

9.4 Applications of Learning Based Algorithms in Anomaly Detection

Several applications of anomaly detection will be discussed in this section.

Intrusion detection: The process of detecting hostile behavior in a computer system is known as intrusion detection. These malicious activities can be in penetrations, break-ins, or other forms of computer abuse. These malicious behaviors or attacks are noteworthy from the standpoint of computer security. Anomaly detection methods can be used in intrusion detection as once the system is compromised, it behaves differently. Sometimes, data has to be explicitly augmented for generalization [30].

Detection of Frauds: The process of detecting illegal activity in commercial organizations such as credit card firms, stock exchanges, mobile phone providers, banks, and insurance agencies is referred to as fraud detection. The malevolent users might be legitimate clients of the company or could be impersonating. Fraud can be detected when a user, as mentioned above, illegally utilize an organization's resources.

Industrial damage detection: Prolonged use and regular wear and tear cause industrial units to deteriorate. To avoid future escalation and costs, such damages must be discovered early. Since it is recorded using various sensors and gathered for analysis, the data in this sector is commonly referred to as sensor data. In order to detect such damages, anomaly detection techniques have been widely used in this field.

Sensor fault detection: IoT sensor networks have recently been a significant area of research, particularly in data processing, because sensor data generated from various wireless sensors have several distinct features. This collected sensor data might contain anomalies that indicate that there may be one or multiple malfunctioning sensors. These anomalies might also indicate that there has been an intrusion in the sensor networks that can be intriguing for data analysts. Blockchain can be used to secure the data [31]. As a result, anomaly detection in sensor networks can identify sensor faults, intrusions, or both. Furthermore, to visualize such high-dimensional sensor data into two-dimensional space, T-SNE is widely used [32].

9.5 Framework

The proposed framework is described in Fig. 9.3. It has mainly three stages: Dataset Collection, Data Pre-Processing, Detection Algorithm Training, and the final evaluation. The TensorFlow [33] and Keras [34] are the deep learning frameworks used for modeling purposes. Below we provide details for each of the stages in brief.

Stage-1: Data Collection: We use the open-source Credit Card Fraudulent Transaction Dataset available on Kaggle [34]. There are 28 features in the form of Principal Component Analysis. The total length of the dataset is 284,807 transactions, where 492 are fraud transactions, and the rest are legitimate. This is a highly imbalanced dataset.

Stage-2: Data Preprocessing: We perform the following data preprocessing tasks:

(1) Standardize the dataset
(2) Split the dataset for training, validation, and testing using stratified sampling.

Stage-3: Algorithm Training: This is the primary step of the whole structure where we train multiple algorithms. The figure shows the AutoEncoder architecture (Fig. 9.4).

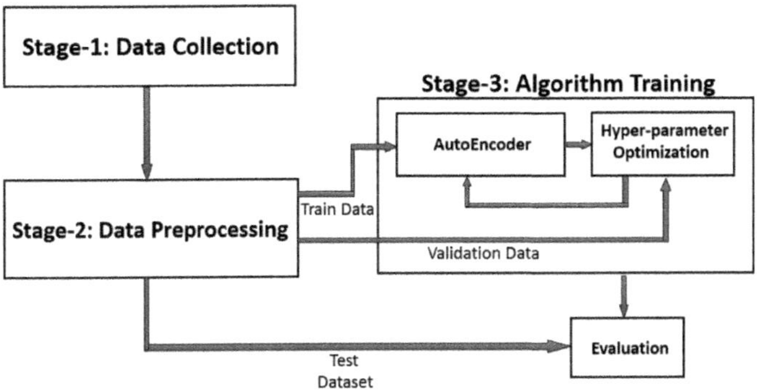

Fig. 9.3 Proposed framework

Fig. 9.4 Model architecture

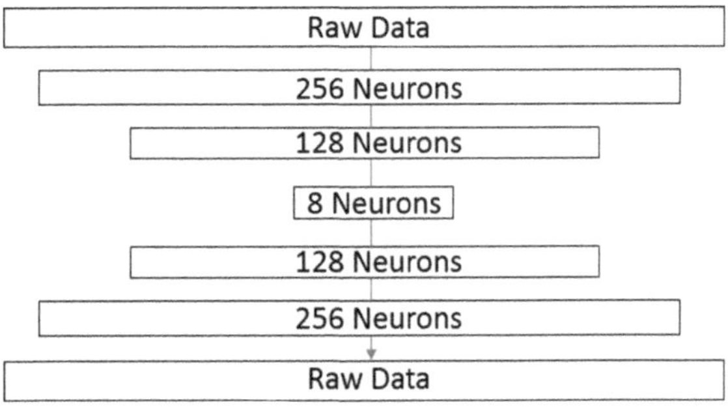

The proposed model is trained with the normal class of data, and thus it learns the patterns of the legitimate transactions. We use Adam optimizer to optimize the weight using gradient descent algorithm with Rectifier Linear Unit (ReLu) as activation function. Next, the batch size is set as 128. Grid Search is used to optimize the hyper parameters of the model. We trained our model with 50 epochs. The proposed model achieved a loss of 0.0008. The model correctly classified the test data as a fraud or legitimate with a 94% detection rate.

9.6 Conclusion

In the Financial market, credit card fraud has increased at an alarming rate. Due to this, cardholders are uncertain about making purchases, and therefore, both the financial organizations and the merchants bear huge losses. Detection of such fraudulent transactions is difficult using traditional supervised machine learning due to the high imbalance nature of labeled data. Thus in this chapter, we discuss several machine learning and deep learning anomaly detection [35–37] algorithm which learn the patterns of the normal class of data [38, 39]. Furthermore, we gave a practical implementation of applying an algorithm to solving the credit card fraud detection problem using a real-world open-source dataset. This approach can be extended by applying the algorithms on cloud based techniques [40–43].

References

1. https://www.iii.org/fact-statistic/facts-statistics-identity-theft-and-cybercrime
2. N.V. Dharwadkar, A.A. Dixit, A.K. Kannur, M.A.B. Kadampur, S. Joshi, Identification of reasons behind infant crying using acoustic signal processing and deep neural network for neonatal intensive care unit. Int. J. Inf. Retr. Res. (IJIRR) **12**(1), 1–17. https://doi.org/10.4018/IJIRR.289576
3. R. Brause, T. Langsdorf, M. Hepp, Neural data mining for credit card fraud detection, in *Proceedings 11th International Conference on Tools with Artificial Intelligence.* (IEEE, 1999), pp. 103–106
4. S. Joshi, H. Upadhyay, L. Lagos, N.S. Akkipeddi, V. Guerra, Machine learning approach for malware detection using random forest classifier on process list data structure, in *Proceedings of the 2nd International Conference on Information System and Data Mining (ICISDM '18)* (Association for Computing Machinery, New York, NY, USA, 2018), pp. 98–102. https://doi.org/10.1145/3206098.3206113
5. A. Srivastava, A. Kundu, S. Sural, A. Majumdar, Credit card fraud detection using hidden Markov model. IEEE Trans. Dependable Secure Comput. **5**(1), 37–48 (2008)
6. T. Bhardwaj, R. Mittal, H. Upadhyay, L. Lagos, Applications of swarm intelligent and deep learning algorithms for image-based cancer recognition, in *Artificial Intelligence in Healthcare* (Springer, Singapore, 2022), pp. 133–150
7. T. Bhardwaj, H. Upadhyay, L. Lagos, Deep learning-based cyber security solutions for smart-city: application and review, in *Artificial Intelligence in Industrial Applications* (Springer, Cham, 2022), pp. 175–192
8. V. Verma, L. Bian, D. Ozecik, S.S. Sirigineedi, A. Leon, Internet-enabled remotely controlled architecture to release water from storage units, in *World Environmental and Water Resources Congress 2021* (pp. 586–592)
9. J.T. Quah, M. Sriganesh, Real-time credit card fraud detection using computational intelligence. Expert Syst. Appl. **35**(4), 1721–1732 (2008)
10. R.J. Bolton, D.J. Hand, Statistical fraud detection: a review. Stat. Sci. **17**(3), 235–255 (2002)
11. T. Bhardwaj, H. Upadhyay, S.C. Sharma, Autonomic resource provisioning framework for service-based cloud applications: a queuing-model based approach, in *2020 10th International Conference on Cloud Computing, Data Science & Engineering (Confluence)* (IEEE, 2020), pp. 605–610
12. T. Bhardwaj, H. Upadhyay, S.C. Sharma, An autonomic resource allocation framework for service-based cloud applications: a proactive approach, in *Soft Computing: Theories and applications* (Springer, Singapore, 2020), pp. 1045–1058
13. T. Bhardwaj, H. Upadhyay, S.C. Sharma, Framework for quality ranking of components in cloud computing: regressive rank, in *2020 10th International Conference on Cloud Computing, Data Science & Engineering (Confluence)* (IEEE, 2020), pp. 598–604
14. S.S. Sirigineedi, J. Soni, H. Upadhyay, Learning-based models to detect runtime phishing activities using urls. In *Proceedings of the 2020 the 4th International Conference on Compute and Data Analysis* (2020), pp. 102–106
15. D. Gangwani, P. Gangwani, Applications of machine learning and artificial intelligence in intelligent transportation system: a review, in *Lecture Notes in Electrical Engineering* (Springer, 2021), pp. 203–216

16. J. Soni, N. Prabakar, J.H. Kim, Prediction of component failures of telepresence robot with temporal data, in *30th Florida Conference on Recent Advances in Robotics* (2017)
17. S.M. Erfani, S. Rajasegarar, S. Karunasekera, C. Leckie, High-dimensional and large-scale anomaly detection using a linear one-class SVM with deep learning. Pattern Recogn. **58**, 121–134 (2016)
18. F.T. Liu, K.M. Ting, Z.H. Zhou, Isolation forest, in *2008 Eighth IEEE International Conference on Data Mining*. IEEE, 2008), pp. 413–422
19. M. Ashrafuzzaman, S. Das, A.A. Jillepalli, Y. Chakhchoukh, F.T. Sheldon, Elliptic envelope based detection of stealthy false data injection attacks in smart grid control systems, in *2020 IEEE Symposium Series on Computational Intelligence (SSCI)*. (IEEE, 2020), pp. 1131–1137
20. R. Todeschini, D. Ballabio, V. Consonni, F. Sahigara, P. Filzmoser, Locally centred Mahalanobis distance: a new distance measure with salient features towards outlier detection. Anal. Chim. Acta **787**, 1–9 (2013)
21. K. Greff, R.K. Srivastava, J. Koutník, B.R. Steunebrink, J. Schmidhuber, LSTM: a search space odyssey. IEEE Trans. Neural Netw. Learn. Syst. **28**(10), 2222–2232 (2016)
22. J. Soni, N. Prabakar, H. Upadhyay, Behavioral analysis of system call sequences using LSTM Seq-Seq, cosine similarity and jaccard similarity for real-time anomaly detection, in *2019 International Conference on Computational Science and Computational Intelligence (CSCI)* (IEEE, 2019), pp. 214–219
23. P. Gangwani, J. Soni, H. Upadhyay, S. Joshi, A deep learning approach for modeling of geothermal energy prediction. Int. J. Comput. Sci. Inf. Secur. **18**(1), 62–65 (2020)
24. J. Soni, N. Prabakar, H. Upadhyay, Comparative analysis of LSTM sequence-sequence and auto encoder for real-time anomaly detection using system call sequences (2019)
25. Z. Huang, W. Xu, K. Yu, Bidirectional LSTM-CRF models for sequence tagging (2015). arXiv: 1508.01991
26. R. Fu, Z. Zhang, L. Li, Using LSTM and GRU neural network methods for traffic flow prediction, in *2016 31st Youth Academic Annual Conference of Chinese Association of Automation (YAC)* (IEEE, 2016), pp. 324–328
27. A. Ng, Sparse autoencoder. CS294A Lect. Notes **72**, 1–19 (2011)
28. J. Soni, N. Prabakar, H. Upadhyay, Feature extraction through deepwalk on weighted graph, in *Proceedings of the 15th International Conference on Data Science (ICDATA'19), Las Vegas, NV* (2019)
29. D. Gangwani, Q. Liang, S. Wang, X. Zhu, An empirical study of deep learning frameworks for melanoma cancer detection using transfer learning and data augmentation. IEEE Int. Conf. Big Knowl. (ICBK) **2021**, 38–45 (2021). https://doi.org/10.1109/ICKG52313.2021.00015
30. P. Gangwani, A. Perez-Pons, T. Bhardwaj, H. Upadhyay, S. Joshi, L. Lagos, Securing environmental IoT data using masked authentication messaging protocol in a DAG-based blockchain: IOTA tangle. Future Internet **13**(12), 312 (2021)
31. J. Soni, N. Prabakar, H. Upadhyay, Visualizing high-dimensional data using t-distributed stochastic neighbor embedding algorithm, in *Principles of Data Science*. (Springer, Cham, 2020), pp. 189–206
32. J.V. Dillon, et al., Tensorflow distributions (2017). arXiv:1711.10604
33. N. Ketkar, Introduction to keras, in *Deep Learning with Python* (Apress, Berkeley, CA, 2017), pp. 97–111
34. https://www.kaggle.com/mlg-ulb/creditcardfraud
35. D.P. Andrea, B. Giacomo, C. Olivier, A. Cesare, B. Gianluca, Credit card fraud detection: a realistic modeling and a novel learning strategy. IEEE Trans. Neural Netw. Learn. Syst. **29**(8), 3784–3797, IEEE (2018)
36. H. Upadhyay, L. Lagos, S. Joshi, A. Szilagyi, Artificial intelligence application to D and D—20492, in *Conference: WM2020: 46. Annual Waste Management Conference, Phoenix, AZ (United States)* (2020). https://www.xcdsystem.com/wmsym/2020/index.html
37. H. Upadhyay, L. Lagos, S. Joshi, A. Abrahao, Big data framework with machine learning for D and D applications—19108, in *Conference: WM2019: 45. Annual Waste Management Conference, Phoenix, AZ (United States)* (2019). https://www.xcdsystem.com/wmsym/2019/index.html

38. S. Joshi, H. Upadhyay, L. Lagos, Deactivation and decommissioning web log analysis using big data technology—15710, in *Conference: WM2015: Annual Waste Management Symposium, Phoenix, AZ (United States)* (2015). http://archive.wmsym.org/2015/index.html
39. H.U.K. Kooper, M. Esoofally, S. Joshi, L. Lagos, Predictive analytics with big data-spark framework. Nuclear Plant J. https://digitaleditions.nuclearplantjournal.com/MA18/29/
40. T. Bhardwaj, C. Reyes, H. Upadhyay, S.C. Sharma, L. Lagos, L, Cloudlet-enabled wireless body area networks (WBANs): a systematic review, architecture, and research directions for QoS improvement. Int. J. Syst. Assur. Eng. Manag. 1–25 (2021)
41. T. Bhardwaj, H. Upadhyay, S.C. Sharma, Autonomic resource allocation mechanism for service-based cloud applications, in *2019 International Conference on Computing, Communication, and Intelligent Systems (ICCCIS)*, pp. 183–187 (2019). https://doi.org/10.1109/ICCCIS48478.2019.8974515
42. T. Bhardwaj, S.C. Sharma, An autonomic resource provisioning framework for efficient data collection in cloudlet-enabled wireless body area networks: a fuzzy-based proactive approach. Soft Comput. **23**, 10361–10383 (2019). https://doi.org/10.1007/s00500-018-3587-x
43. T. Bhardwaj, S.C. Sharma, Cloud-WBAN: an experimental framework for cloud-enabled wireless body area network with efficient virtual resource utilization. Sustain. Comput.: Inform. Syst. **20**, 14–33 (2018)

Printed by Printforce, the Netherlands